D1376111

Basic Maths

FOR

DUMMIES®

Basic Maths
FOR
DUMMIES®

by Colin Beveridge

A John Wiley and Sons, Ltd, Publication

Basic Maths For Dummies®

Published by
John Wiley & Sons, Ltd
The Atrium
Southern Gate
Chichester
West Sussex
PO19 8SQ
England

Email (for orders and customer service enquires): cs-books@wiley.co.uk

Visit our home page on www.wiley.com

For general information on our other products and services, please contact our Customer Care Department within the U.S. at 877-762-2974, outside the U.S. at 317-572-3993, or fax 317-572-4002.

For technical support, please visit www.wiley.com/techsupport.

Wiley also publishes its books in a variety of electronic formats and by print-on-demand. Some content that appears in standard print versions of this book may not be available in other formats. For more information about Wiley products, visit us at www.wiley.com.

British Library Cataloguing in Publication Data: A catalogue record for this book is available from the British Library

ISBN: 978-1-119-97452-9 (paperback), 978-1-119-97561-8 (ebook), 978-1-119-97562-5 (ebook), 978-1-119-97563-2 (ebook)

Printed and bound in Great Britain by TJ International, Padstow, Cornwall

10 9 8 7 6 5

About the Author

Colin Beveridge is a maths confidence coach for Flying Colours Maths and co-author of the *Little Algebra Book*.

He holds a PhD in Mathematics from the University of St Andrews and worked for several years on NASA's Living With A Star project at Montana State University, where he came up with an equation which is named after him. It's used to help save the world from being destroyed by solar flares. So far so good.

He became tired of the glamour of academia and returned to the UK to concentrate on helping students come to terms with maths and show that not all mathematicians are boring nerds; some are exciting, relatively well-adjusted nerds.

Colin lives in Poole, Dorset with an espresso pot, several guitars and nothing to prove. Feel free to visit his website at www.flyingcoloursmaths.co.uk or follow him on Twitter at www.twitter.com/icecolbeveridge.

Dedication

For the teachers who taught me most of what I know: Brian Rodrigues, DJ Rowley, Dana Longcope and Naomi Dunford.

And for the students who taught me most of the rest.

Author's Acknowledgements

I'm very grateful to the team at Dummies Towers for their work and guidance in making this book awesome – particularly my editors Rachael Chilvers and Mike Baker.

The writing was largely fuelled by the Little Red Roaster coffee shop in Parkstone, and inspired by the students who helped me develop the ideas and make them simple enough to understand – extra-special thanks to Tain Duncan, Ethan Oak and Dale Bannister. LaVonne Ellis, Lisa Valuyskaya and Ryah Albatros from Customer Love all went above and beyond in getting me to just write the blasted thing.

It wouldn't have been written at all without the work my brother Stuart and his colleagues at The Chase did on the *Little Algebra Book*, or the unfaltering support of my parents – Ken Beveridge and Linda Hendren – and the tireless encouragement of Melissa Day.

Publisher's Acknowledgements

We're proud of this book; please send us your comments through our Dummies online registration form located at www.dummies.com/register/.

Some of the people who helped bring this book to market include the following:

Commissioning, Editorial, and Media Development

Project Editor: Rachael Chilvers

Commissioning Editor: Mike Baker

Assistant Editor: Ben Kemble

Development Editor: Colette Holden

Technical Editors: Samuel Harrison, Vincent Kwasnica

Proofreader: Jamie Brind

Production Manager: Daniel Mersey

Publisher: David Palmer

Cover Photo: © Shutterstock / Robert Spriggs

Cartoons: Ed McLachlan

Composition Services

Project Coordinator: Kristie Rees

Layout and Graphics: Corrie Socolovitch, Christin Swinford

Proofreader: Lauren Mandelbaum

Indexer: Becky Hornyak

Publishing and Editorial for Consumer Dummies

Kathleen Nebenhaus, Vice President and Executive Publisher

Kristin Ferguson-Wagstaffe, Product Development Director

Ensley Eikenburg, Associate Publisher, Travel

Kelly Regan, Editorial Director, Travel

Publishing for Technology Dummies

Andy Cummings, Vice President and Publisher

Composition Services

Debbie Stailey, Director of Composition Services

Contents at a Glance

Table of Contents

Introduction

*H*i! I'm Colin, and I want to change the world.

I live in a world where, when I say 'I'm a maths tutor,' people say to me 'Oh, maths . . . I was never any good at that,' or 'I haven't used maths since I left school.'

I live in a world where I have to bite my tongue rather than say 'I don't mind that you think maths is hard, but I am saddened that you're proud of this,' or 'Are you sure about that? I bet you used maths six times before breakfast this morning.'

I want to live in a world where everyone is okay at maths. Not a world full of Einsteins, not a world full of geeks – just a world where having a solid level of maths is as natural to everyone as having a solid level of reading and talking.

This book is part of my vision – and I'm delighted that you've picked it up. You've taken the first step to being a bigger part of my ideal world, and I want to do everything I can to help you become as good at maths as you want to be.

I want to show you that maths makes sense, most of the time, and that you use maths countless times a day, often when you don't even realise it. I want to show you that after you break problems down into smaller steps, those problems are so much more doable. I want to help you stop feeling stupid or afraid or troubled by maths. I know you aren't stupid: you just haven't got the hang of maths yet.

Most of all, I want to hear how you get on. The best way to catch me is on twitter (@icecolbeveridge) – I can't promise I'll get back to you straight away, but I promise I'll read and reply if I possibly can.

About This Book

In this book, I try to help you understand not only *how* to do the maths you need, but *why* you do the maths in a certain way. I show you maths isn't some mystical language of squiggles but instead is a concise and efficient way to communicate. One of the nice things about maths is that it changes very little from country to country. I studied maths in France for a year and was surprised that most of the words, even in the ridiculously advanced maths they thought I could do, were either basic French vocab or very similar to the English words. The sums were exactly the same. (I still couldn't do them, but that's a different story.)

Now, I'm pretty good at maths. I've been a full-time maths tutor since 2008. Before that I worked on a NASA project in the USA. I have an equation named after me. I know my stuff.

But – and this is a big 'but' – I know that being a good mathematician isn't the same as being a good maths teacher. I'm lucky to have worked with enough people at the stage you are now – smart and interested, but needing help to understand – that I can break down maths into smaller, simpler parts that I hope you'll understand.

Among other things, I cover the following in this book:

- ✔ Keeping calm instead of stressing about maths.
- ✔ Solving regular arithmetic problems – adding and taking away, multiplying and dividing.
- ✔ Rounding off and estimating your answers.
- ✔ Dealing with decimals, fractions, percentages and ratios.
- ✔ Messing about with measures of time, money, weight and temperature.
- ✔ Understanding shapes – how you measure them and how you move them around.
- ✔ Grappling with graphs – both reading and drawing.
- ✔ Summing up statistics, including averages and probabilities.

How could that not be fun?

This book is based on the UK Adult Numeracy Core Curriculum, from Entry Level 3 through to Level 2. Whether or not a Level 2 numeracy qualification is equivalent to an A–C pass at GCSE is a murky area I don't want to muddy further, but I reckon they're roughly the same level in difficulty, although the numeracy curriculum covers slightly fewer topics.

So, that means this book may help you do pretty well at GCSE level but won't cover all of the topics involved – particularly algebra, which this book barely touches. If you read this book cover to cover and understand everything, you should ace any UK Adult Numeracy test thrown at you. Although I've based this book on the core curriculum, I sometimes dip into topics in a little more detail than needed. I also cover a few areas from a slightly earlier level in the curriculum if I reckon you may find the particular subject hard.

Whether you're studying for a numeracy qualification or a GCSE, or just want to brush up on your basic maths skills, this book has what you need. Best of all, the book follows the *For Dummies* format. Divided into easy-to-follow parts, the book serves as both your reference and your troubleshooting guide.

Conventions Used in This Book

I keep the conventions to a minimum in this book. Here are the ones I use:

- ✔ I use *italics* for emphasis or to highlight new words or phrases.
- ✔ **Boldfaced** text indicates key words in bulleted lists or the key steps of action lists.
- ✔ Monotype font is used for internet and email addresses.

What You're Not to Read

This book is designed to be an easy-access reference guide to basic maths. I cover each subject in its entirety in individual chapters, and the information doesn't depend on what comes before or after. This means you can jump around the book to the subjects you want to focus on and skip those you feel comfortable with already or just aren't interested in.

If you feel like you're starting from scratch, I strongly recommend you peruse the whole book to get a solid idea of all that's involved. If you already have a decent maths background, you probably want to focus on the areas you find are relatively weak for you – but you may also find some of the insights in other areas help to shore up your maths skills.

No matter what your background, you can skip paragraphs marked with the Technical Stuff icon without giving up an understanding of the primary subject. Also, sidebars supplement the primary text – you can skip them without missing the main point.

Foolish Assumptions

Making assumptions is always a risky business, but knowing where I'm coming from may put you at ease. So, in writing this book, I assume that:

- ✔ You know how to count and are familiar with the symbols for the numbers.

- ✔ You understand the idea of money and changing a banknote for an equivalent value of coins.

- ✔ You know what some basic shapes look like.

- ✔ You're prepared to think fairly hard about maths and want either to pass a numeracy qualification or to simply brush up on your maths skills.

How This Book Is Organised

Like all *For Dummies* books, *Basic Maths For Dummies* is a reference and each topic is allotted its own part in the book. Within each part are individual chapters relating specifically to the topic in question.

Part 1: Whole Numbers: The Building Blocks of Maths

If you want to be good at maths – and I presume you do, otherwise you'd have picked up a different book – being able to do three things really well is helpful:

- ✔ **Stay calm:** Maths can be hard enough when your mind isn't complaining about how impossible it is and trying to sabotage your efforts. I give you tips on how to keep that under control and set yourself up for success.

- ✔ **Do sums on paper:** As a mathematician, I think this is the least important of the three for being good at maths. Unfortunately, the people who write exams disagree with me, so I walk you carefully through methods for adding, taking away, multiplying and dividing, and I try to show you *why* the sums work.

- ✔ **Work out rough answers:** Being able to give a ballpark answer quickly is far more important to me than being able to work out a huge long-division sum. I show you how to work out a rough answer without giving yourself a headache.

Part II: Parts of the Whole

Stand back, everybody – I'm going to use the F-word. A word some people would like to see banned from books in public libraries and never have to hear on the TV. That's right, I'm talking about fractions. *Gasp! I can't believe he wrote that!*

I'm here to tell you that there's nothing dirty about fractions, even improper ones. There's a lot of misinformation out there about fractions, and some teachers find talking about fractions very difficult without getting embarrassed. But don't worry – in Part II I try to answer all of your questions about fractions in a frank and easy-to-understand manner. I also cover their close friends, decimals, percentages, ratios and proportion, which are all versions of the same thing.

I also introduce you to the Table of Joy – an easy way to work with percentages, ratios and literally dozens of other topics. I use this table throughout the book. In fact, the Table of Joy is probably the most useful thing I know.

Part III: Sizing Up Weights, Shapes and Measures

The third part of this book is about applying your maths knowledge to real-life things – generally things that you measure.

Some of these concepts are perfectly familiar – you've probably worked with time and money since you were old enough to throw a Monopoly board across the room. However, there are places you need to be careful – and this book gives you a few extra tips and tricks to pick up in those areas.

Some of the measuring concepts are a bit trickier. I look at the different ways to measure weight and temperature and show you some of their many applications.

I also look at size and shapes – again, there are different ways of measuring these and many facets of shape to play with.

Part IV: Statistically Speaking

Statistics has a reputation for being boring and difficult. For a long while, I bought into that story too – but then I started using statistics and applying it to something I cared about. Suddenly, I was drawing graphs that helped me understand my project, working out statistics that told me what was going on and making predictions based on probabilities . . . and I was hooked.

I can't promise you'll find statistics as exciting as I do, but I do my best to make the topic interesting. I cover the ins and outs and ups and downs of graphs and tables, how to interpret them and how to draw them; I look at averages; and I dip a toe into the murky and controversial world of probability.

Part V: The Part of Tens

All *For Dummies* books finish with The Part of Tens, a bunch of lists full of practical tips to help you manage the material in the rest of the book.

I run you through ways of calming down and some ideas for remembering your number facts. I show you some of the booby-trap questions examiners may set, and I offer some exam-technique tips so you can get in there and ace it. Good luck!

Icons Used in This Book

Here are the icons I use to draw your attention to particularly noteworthy paragraphs:

Theories are fine, but anything marked with a Tip icon in this book tells you something practical to help you get to the right answer. These are the tricks of the mathematical trade.

Paragraphs marked with the Remember icon contain the key takeaways from the book and the essence of each subject.

The Warning icon highlights errors and mistakes that can cost you marks or your sanity, or both.

You can skip anything marked with the Technical Stuff icon without missing out on the main message, but you may find the information useful for a deeper understanding of the subject.

Where to Go from Here

This book is set up so you can jump right into the topics that interest you. If you feel like an absolute beginner in maths, I recommend you read Parts I and II to build a foundation for the other topics. If you're pretty comfortable with the mechanics of maths, use the table of contents and index to find the subject you have questions about right now. This book is a reference – keep it with your maths kit and turn to it whenever you have a question about maths.

Part I
Whole Numbers: The Building Blocks of Maths

THE FOUR HORSEMEN OF THE ARITHMETICS

In this part . . .

If you can count, you can do maths.

It's helpful to build up shortcuts to make maths easier, though – and that's what this part is all about: making maths easier. I show you how to stay calm and focused (and shut up the little voices telling you you can't do maths) and then help you figure out how to add, take away, multiply and divide whole numbers – the sums all of the others are based on.

You need to be able to see if your answer looks right: to do that, you need to be able to round off and to estimate so that you don't say something daft like 'The Eiffel Tower is four centimetres tall'.

Chapter 1

Getting Started

*B*efore you read any more of this book, take a big, deep breath. I know what taking on something difficult or frightening feels like – I feel just the same about dance classes, and I still have to steel myself a bit when I go into a supermarket.

I start this chapter by saying thanks – thanks for giving maths a try and thanks for listening to me. I'm not the kind of maths teacher who wears tweed jackets with leather patches and yells at you when you don't pick up on his mumbles straight away. I want to help you get past the fear and the mind blanks and show you not just that you *can* do maths well, but that you *already* do maths well and can use that base to build upon. I show you how, with a bit of work, you can master the bits and pieces of maths you don't have down to a tee. You're smart. I believe in you.

Perhaps you find the maths you do in day-to-day life so easy you don't even notice you're doing sums. I spend some time in this chapter showing you what you already know and then introduce the topics I cover in the rest of the book.

You're Already Good at Maths

Put your hand up if you've ever said something like 'I'm no good at maths.' I promise I won't yell at you. Now imagine saying 'I'm no good at talking' or 'I'm no good at walking.' Those things may be true at times – I get tongue-tied once in a while, and I've been known to trip over invisible objects – but most of the time my mumbling and stumbling are perfectly adequate to get by. I bet the same thing applies with your maths. Maybe you freeze up when you see a fraction or just nod and smile politely when someone shows you a pie chart. This doesn't mean you're bad at maths, just that you trip up once in a while.

If you can shift your thoughts on maths from 'I'm no good at this' to 'I'm still getting to grips with this', you'll create a self-fulfilling prophecy and begin to understand maths.

Part of the problem may be that you don't realise how much of what you do every day involves doing maths in your head. You may not *think* you're doing maths when you judge whether to cross the road on a red light, but your brain is really doing a series of complex calculations and asking questions such as:

- ✔ How fast is that bus going, and how far away is it? How long will the bus take to get here?

- ✔ How wide is the road, and how long will it take for me to get across?

- ✔ What's the probability of that driver slowing down to avoid me if I'm in the road?

- ✔ How badly do I want to avoid being honked at or run over?

- ✔ What are the survival and recovery rates for my local hospital?

- ✔ How soon do I need to be where I'm going?

- ✔ How much time will crossing now save over waiting for the light to change?

You do all of these calculations – very roughly – in your head, without a calculator, and without freezing up and saying 'I'm no good at maths.' If you regularly got any of those sums wrong – the speed–distance–time analysis, the probability or the game theory – you'd be reading this in hospital and trying to figure out what the jagged line graph at the end of the bed means. (Turn to Chapter 16 if this really is the case – and get well soon!)

So before you cross the road on your way to work, you solve as many as six 'impossible' sums in your head, maybe before you've even had breakfast.

Your First Homework Assignment

I'm not a big one for setting homework, but I'm going to ask you to do one thing for me (and, more importantly, for yourself): if you ever find yourself in a situation where you feel like saying 'I'm no good at maths', catch yourself and say something else. Try 'I used to struggle with maths, but I'm discovering that maths is easier than I thought', or 'I'm fine with day-to-day maths', or 'I really recommend *Basic Maths For Dummies*: this book turned me into a mathematical genius.'

Although mathematicians traditionally wear rubbish clothes, thick glasses and a bad comb-over, this fashion isn't compulsory. The tweed generation is dying out, and most of the maths geeks I know are now just a bit scruffy. So, don't worry: being good at maths won't turn you into a fashion disaster with no friends.

I appreciate my homework assignment is tremendously difficult – asking you to change your entire way of thinking is a big ask. To assist you I enlist the help of an elastic band and ask you to treat yourself with something I call Dunford Therapy, after the genius who told me about it:

1. **Find an elastic band big enough to go around your wrist comfortably.**

   Put the elastic band around one of your wrists – either one, it doesn't matter.

2. **Every time you catch yourself thinking anything along the lines of 'I'm no good at maths', snap the elastic band *really hard* against the bony bit of your wrist.**

   This will hurt. That's the idea.

3. **After you catch yourself a few times, your brain will start to rewire itself to avoid thinking such filthy and disgusting thoughts, and you'll find yourself capable of extraordinary feats of mathematics.**

If you have particularly fragile wrists or any inkling that you might do yourself more damage with an elastic band than swearing and shaking your hand in pain, don't use Dunford Therapy. The elastic band is supposed to hurt just enough to help you change your way of thinking, not to injure you.

Getting the odd maths sum wrong doesn't mean you are stupid – far from it in fact, because you're immediately and obviously smarter than someone who doesn't even try the sum.

Talking Yourself Up

Encouraging yourself is a recurring theme in this book – the more you give yourself credit for the things you *can* do, the easier the things you're still working on become. Be sensible about things: don't rush to the library and check out the *Journal of Differential Equations* (at least, not until you've bought and devoured *Differential Equations For Dummies*). But when you see something that's a bit tricky-looking, try to avoid saying 'I can't do that' or 'I haven't been taught that' as a response. Maybe say 'I can't do that *yet*' or 'I need to do some work on this.' Better still, say 'What would I need to find out to be able to solve this?'

Chapter 2 is all about ways to build your confidence and set yourself up to get on top of your maths studies quickly, effectively, and with a great big goofy grin. Best of all, Dunford Therapy isn't part of Chapter 2.

Whole Numbers: Party Time!

Everyone likes parties. Balloons! Silly hats! Cheese-and-pineapple sticks arranged in a potato to look like a hedgehog! But these things don't spring into existence on their own. If you want to plan a party, you may need to put your maths skills to work to make sure you prepare enough vol-au-vents for everyone.

Maybe you want to bake a cake for 12 people coming to celebrate your birthday. But disaster! Your recipe book only has a recipe for four people. What can you possibly do?

I'm sure you can come up with a few solutions. I've also got a few ideas, which I explain here in excruciating detail:

- ✓ **Let people go hungry:** You have 12 guests and only enough cake for four. How many will have to forgo your delicious Victoria sponge? Twelve people take away four lucky cake-eaters leaves eight guests, who probably need to go on a diet anyway.

- ✓ **Make extra cakes:** One cake feeds four people and you want to feed 12. How many cakes do you need? Twelve people divided by four per cake gives you three cakes.

- ✓ **Cut your slices into smaller pieces:** If you cut four slices each into three smaller bits, you have 4 times 3 equals 12.

- ✓ **Make a bigger cake:** This is the kind of approach that you typically get asked about in an exam. You need to figure out how much bigger to make the cake – just like before, $12 \div 4 = 3$ times as big. To make the cake three times bigger, you multiply all of the ingredients in the recipe by three.

My suggestion above is a bit of a 'don't try this at home' moment: although the last option is the most 'mathsy', it may not work out quite as well in real life. Unless the recipe in your cookery book gives instructions on how to adjust the cooking time of your humungous new cake, the physics of cake-baking may conspire against you and leave you with something inedible. Try my idea if you like, but don't blame me if your cake doesn't rise.

Forgive me if you already knew how to do all of that. That's actually a good sign. The point wasn't to bamboozle you with tricky maths but to say that sometimes you do maths without even thinking about what you're doing.

One of the points from my example above is to think about which sum is appropriate for each idea, so you can adapt the concept to different situations. What if your cake recipe serves six people? What if you're expecting 48 guests? What if the recipe is for casserole instead of cake?

In Part I of this book I look at exactly this kind of question. What kind of sum is the right one to do? How can you figure out roughly what the answer should be? How do you work out the arithmetic to get a precise answer? I look at the 'big four' operations – adding, taking away, multiplying and dividing – along with estimating and rounding to get rough answers.

Parts of the Whole: Fractions, Decimals, Percentages and More

Public speaking . . . death . . . spiders . . . fractions. Are you scared? *Boo!* Are you scared now?

I understand. Seeing how whole numbers fit together is relatively easy, but then suddenly the evil maths guys start throwing *fractions* at you – and then things aren't so intuitive. Fractions (at least, proper fractions) are just numbers that are smaller than whole numbers – they follow the same rules as regular numbers but sometimes need a bit of adjusting before you can apply them to everyday situations.

I have two main aims in this section: to show you that fractions, decimals, percentages and ratios are nothing like as fearsome as you may believe; and to show you that fractions, decimals, percentages and ratios are all different ways of writing the same thing – therefore, if you understand one of them, you can understand all of them.

I won't promise that you'll emerge from this section deeply in love with fractions, but I hope I can help you make peace with fractions so you can work through the questions likely to come up in exams and in real life.

Mmmm, pizza! Everyday fractions

You use fractions and decimals in real life all the time – any time you slice a pizza into smaller bits . . . any time you say you'll be somewhere at quarter past six . . . any time you say or read the price of a product in the supermarket and, in fact, any time at all when you use money.

A fraction is really just two numbers, one on top of the other, that describe an amount (usually, anyway) between zero and one. A fraction is a part of a whole one. The bottom number tells you how finely you've divided the whole thing (the bigger the number, the finer or smaller the 'slice') and the top number tells you how many slices you have.

For example, think about a quarter of an hour. A quarter is written as ¼: the 4 says 'Split your hour into four equal bits', and the 1 says 'Then think about one of the bits.' A quarter of an hour is a whole hour (or 60 minutes) divided into four parts, making 15 minutes. Three-quarters of an hour (¾) is three times as long: 45 minutes.

You already use decimals all the time as well. When you write down an amount of money using pounds and pence, you use a decimal point to show where the whole number (of pounds) ends and where the parts of a pound (pence) begin. If you look at your phone bill or your shopping receipt, you see decimal points all over the place. Don't be afraid of decimals: as far as you're concerned, decimal points are just dots in a number that you can leave in place and otherwise ignore. For example, you work out a sum like $5.34 \div 2$ (with a dot) in exactly the same as the way you work out $534 \div 2$ (without a dot) – the only difference is that you have to remember to put the dot back in, in the same place, when you finish the sum.

Percentages are easier than you think: Introducing the Table of Joy

What if I told you I had a simple, reliable method for working out the sums you need to do in somewhere between a quarter and a half of questions in a typical numeracy exam? Such a method exists – the Table of Joy. I go into serious detail about this table in Chapter 8, but I also dot it about here and there in other chapters.

You can use the Table of Joy in all of the following topics:

- **Converting imperial to metric units:** Working in either direction, and finding the conversion rate.
- **Currency conversion:** Converting to and from any currency, and working out the exchange rate.

- **Finding a fraction of a number:** Without making you cry.

- **Percentages:** Both regular and reverse percentages.

- **Pie charts:** How big a slice should be, the value a slice represents, and what the total value of the slices in the chart should be.

- **Ratios:** Pretty much any ratio sum you can imagine, and more besides.

- **Recipe scaling:** How much you need to adjust your recipe by, and how many people it now feeds.

- **Scale drawing:** Finding the size of the real thing, or the sketch, or the scale.

- **Speed/distance/time questions:** And pretty much anything you could possibly want to do (at least, that isn't A-level or harder).

Those are just the topics I can think of off the top of my head that are in the numeracy curriculum. You can also use the Table of Joy for things like stratified surveys, histograms, density, gradients, circle theorems and trigonometry.

The idea of the Table of Joy is simple: write down the information you need to use in a labelled table, and do a simple sum to work out the answer to your question. Follow these steps to use the Table of Joy:

1. **Draw out a noughts-and-crosses grid, with squares big enough to label.**

2. **Put the *units* of what you're dealing with in the top-middle and top-right squares.**

   For example, if you want to convert currencies, your units may be 'pounds' and 'dollars'. If you want to work out a sale percentage, your units may be 'pounds' and 'per cent'.

3. **Put the *contexts* of the information you have down the left side.**

   Again, with currencies, you may have 'exchange rate' and 'money changed'. With percentages, you may have 'full price' and 'sale price'.

   Each time the Table of Joy comes up in this book I show you how to label the relevant table, but after a while you'll probably do it instinctively.

4. **Put the relevant numbers in the correct cells, with reference to the labels.**

   For example, 100 per cent is the same as the full price, so 100 goes in the square with 'per cent' at the top and 'full price' at the side.

5. **Put a question mark in the remaining square, and write out the Table of Joy sum.**

   In the Table of Joy you always have three numbers and then work out the fourth.

The sum is the other number in the same row as the question mark, multiplied by the other number in the same column as the question mark, divided by the remaining number.

6. **After you work out the sum, you have your answer.**

This may seem like a lot of work, but after you get into the routine of using the Table of Joy, you'll work out your sums quite quickly. In Figure 1-1 I show how to create an example Table of Joy to answer the following question:

£1 is worth $1.50. I want to buy trainers on sale in America for $75. How much is that in pounds?

Don't worry if the calculation in the Table of Joy looks tricky. In Chapter 7 I take you through decimal sums in detail. I just want to show you here how easily you can figure out what sum you need to do.

Figure 1-1:
The steps of the Table of Joy.
(a) Draw a big noughts-and-crosses grid.
(b) Label the rows and columns.
(c) Fill in the numbers.
(d) Do the Table of Joy sum. The answer is £50.

a.

b.

	£	$
Rate		
Price		

c.

	£	$
Rate	1	1.50
Price	?	75

d.

$$\frac{1 \times 75}{1.50}$$ $75 \div 1.5 = 50$

Sizing Up Time, Weights, Measures and Shapes

I bet you're perfectly comfortable with at least one of the following topics: telling the time, taking a temperature, weighing yourself or other objects, counting money, measuring distances, or playing with shapes. You may even be comfortable with all of them.

Even if you just feel okay with one of these topics, you can build on your knowledge with the other subjects. For example, if you know how to use a thermometer, you can use exactly the same skills to read a scale or a ruler.

This is all useful, day-to-day stuff – and the reason I'm confident you know something about these topics is that they're all around us, all the time. On the way to the supermarket, you may check the bus timetable to see when you need to leave or the weather forecast to see whether you need to wrap up warm. At the shop, you weigh your bananas to see how much they cost, decide between 5 and 10 metres of tinfoil, and then pay the bill at the check-out – before packing the whole lot into your car and making sure it fits nicely.

All of that is maths. And I bet you do most of it without really thinking. Similarly, a lot of students freeze up when I ask them to work out the change from £20 on paper, but given Monopoly money they do the same sum without any trouble. All I want you to do is make sure you link your everyday experiences with the numbers you juggle on the page.

Weights and measures you already know

You have four topics to master that are to do with measuring things other than distance:

- ✔ **Time:** You need to be able to work with a clock, read timetables, fill in timesheets and work with speed – which, after you appreciate the pitfalls, are all pretty easy.

- ✔ **Money:** You're probably already familiar with money sums. You need to be able to do regular arithmetic with money and change one currency into another (the Table of Joy can help).

- ✔ **Weight:** Even if you don't use scales regularly, you've probably seen somebody else use them. I take you through using and reading the various types of scales, and I show you how to convert between different units of weight.

- ✔ **Temperature:** You're probably quite happy with most aspects of temperature, although I do introduce a few tricky bits that you may not get straight away – converting between temperature scales and using negative numbers are two areas where some people end up scratching their heads. Don't worry: I take things slowly.

Getting yourself into shape

You need to understand how to deal with lengths, areas, volumes and shapes. Some people find visualising shapes really easy – if you're one of those people, you'll find the shape chapters pretty straightforward. If not, don't worry – I explain things as simply as possible.

The shape topics are split into two groups: measuring (where I talk about how big a shape is) and the actual shapes (where I help you with angles, symmetry and so on). I also have a quick look at maps and plans. You can measure an object in several different ways. For example, you can measure how tall or wide or deep the object is (length), how much floor space it takes up or how much paper you need to cover it up (area), and how much room space it takes up or how much stuff it holds (volume).

Statistically Speaking

If mathematicians have a bad reputation, then statisticians have it ten times worse (on average). I'd like to let it be known that statisticians' glasses are no thicker, nor their elbows more leather-patched, than those of their mathematical counterparts. Some statisticians – it is alleged – are well-adjusted members of society, although evidence is scant. Understanding graphs and tables, and being able to deal with averages and probability, will not turn you into Statto. In fact, you'll be in a much better place to deal with the statistics that life bombards you with all the time, in the news and maybe at work. You don't need to tell anyone you're studying stats – we can keep it our little secret.

Why bother with charts and tables?

In my back room I have a shoebox full of receipts, bills, statements and handwritten notes saying things like '8.75 on curry' and 'Class with Jenny, £35'. This box is a shambles of an accounting system, and if anyone wants to exchange a few hours of sorting it out for a few hours of maths tuition, please get in touch with me.

But if I want to understand my financial position better, I don't want a shoebox of randomly arranged bits of paper. I want my numbers neatly arranged on a few pages of paper or – better yet – in a graph so I can see at a glance how long I need to work before I pay off my loans and can afford a holiday.

The strength of tables and graphs is that they take a mess of numbers and make them tidier or easier to understand, or both. At the start of Part IV, I show you how to read and make tables and graphs, and help you see which is best to use in which situation.

The man in the middle: Describing data

Another way of tidying up certain data is to describe the data with a *statistic* – a number that tells you something about the data. Examples of a statistic are 'the biggest value' and 'the number of numbers in the data set'. In Chapter 18, I show you how to use four of the most common statistics.

The *mode* is the most common number in a data set and the *median* is the most moderate (half of the numbers are bigger and half of them are smaller). The *mean* is probably what you think of as the average: you get the mean when you divide the total up evenly. The *range* is a measure of how spread out the data are.

What are the chances?

I finish Part IV by looking at probability – a measure of how much you think something will or won't happen. For example, you're more likely to throw a double six on a pair of dice than you are to pull the ace of diamonds out of a pack of shuffled cards. The question is: how do I know that?

I look at what probability means (as best I can – philosophy is a slippery slope) and how you can work out how likely some events are by using straightforward sums.

The Tools You Need

You don't need much stuff to get going with *Basic Maths For Dummies* – you can do an awful lot with just a pen and paper. But you may want to pick up a few extra bits and pieces along the way. Here are the things I strongly recommend you buy the next time you're in a stationery shop or online:

The Adult Numeracy exam

This bit is about registering to take an Adult Numeracy exam. If you study maths at an adult education centre, you don't need to worry about this section: your centre will probably enrol you for the test. Likewise, if you don't plan to take an exam, skip this bit.

Registering for the test

If you want to register for Qualified Teacher Status tests, go to www.tda.gov.uk/trainee-teacher/qts-skills-tests/registering-booking.aspx and register and book through their website. As far as I can see, this is a well-designed site that makes the process easy and quick.

If you want to register for the regular numeracy exam, go to www.move-on.org.uk/findatestcentre.asp. Tell the website where you are and it lists the nearest test centres to you. Alternatively, look for 'Further Education' in the Yellow Pages and make contact with your local centre directly – staff at the centre should tell you what you need to do to sit the test.

What to expect in the test

Adult numeracy tests are generally performed 'on screen'. You go to a test centre and sit at a computer, and the questions come up on the screen. You answer the questions by clicking the mouse on (I hope) the right answer or by pressing the appropriate letter.

The website www.move-on.org.uk has sample tests you can try on your own computer to get used to how they work. I recommend trying them out.

After you finish the test, whether practice or for real, you receive immediate feedback about how you've done.

Look back over your practice tests and see where you went wrong – these are great areas to revise first.

Good luck!

✔ **A geometry set:** A ruler, a set of compasses and a protractor are all useful – as is some of the other stuff that's usually bundled in with those sets . . . although I'm pretty sure no one has used a stencil since about 1994. Some of the processes you need to be able to do to pass the numeracy exam require something from the geometry set, so maybe make this the first bit of equipment you buy.

✔ **A decent calculator:** I recommend the Casio FX-85 – the one with a round button just under the screen, especially if you plan to study more maths after you have the basic stuff under your belt. For this book though, you can do everything on paper; the calculator is really just to check your answer and maybe save some time – any calculator that works will do the trick!

✔ **A notebook:** I find having a single place to keep track of all the things I want to remember makes it easier to look them up when I want to remember them – otherwise you end up scrambling through reams of paper to find the brilliant idea you had three weeks ago but can't remember now.

If you don't have some of this stuff, don't use that as an excuse not to get started! You can make a start on one of the chapters that doesn't need any equipment. For example, in Chapter 10, where I talk about time, you don't need anything more than a pen and paper.

Chapter 2

Setting Yourself Up for Success

・・・

In This Chapter

▶ Setting up an inviting workspace

▶ Staying motivated

▶ Getting your head on straight

▶ Preparing for – and acing – your test

・・・

Don't worry if you're not looking forward to studying (I know that feeling!) – you can make things easier for yourself. You don't have to enjoy learning maths, but I won't tell anyone if you suddenly find yourself thinking 'This isn't so bad after all.' In this chapter I show you how you can set yourself up so that maths is a bit less intimidating, a lot less stressful and (whisper it) more enjoyable. Like most things, success boils down to preparation and positivity – in this chapter, I help you on both of these fronts.

I start off with a few ideas about setting up somewhere to work where you can be comfortable and concentrate with as few distractions as possible, and where everything you need is to hand.

After you set up your perfect workspace, you need time to relax a bit. When you're comfortable, calm and ready to work, learning becomes a million times easier. (Yes, a million times. I checked.) I give you some top tips about calming yourself down and shutting off the little voice that says 'can't' – because when you start believing you can do it, you're halfway there.

Elsewhere in this chapter I delve into the nitty-gritty of preparing for exams. I can describe as many revision techniques as there are people taking exams, but it would be silly to try to cover all of them in this chapter. Instead, I pick out the preparation techniques that my students and I find helpful. Don't try and use all of them at once – but try a few and see what works for you. I also cover the exam itself in this chapter. I provide a few techniques to control your nerves on test day and a few approaches for taking exams so you can squeeze every last mark out of the paper.

If you picked up this book simply to improve your maths skills, feel free to skip the test-taking sections. Likewise, if you'd rather just get on with the maths, flip ahead to one of the other chapters – I won't be offended.

Getting Properly Equipped

I don't like running. Actually, that's not quite true. I don't mind the running part. But I don't like going outside in the cold and wet. I can make myself more likely to go running – and reap its benefits – if I prepare properly. I put my trainers and kit beside my bed so I see them as soon as I get up. I plan a route that takes me through my favourite parts of town. And I promise myself the treat of a good cup of coffee when I get home.

The key is to set myself up properly. I keep the things I know I'll need handy, I make the task as enjoyable as possible, and I reward myself for completing the job.

You can do the same for your maths. By keeping your equipment somewhere convenient, making your workspace comfortable and inviting, and rewarding yourself after each maths session, I bet you soon find studying isn't quite such a chore as you thought.

Tools of the trade

I've been told (by mathematicians) that maths is the second-cheapest subject to study, because all you really need is paper, pencils and a waste-paper bin. Philosophy is a little cheaper because you don't need the bin.

You can do a lot of maths without any expensive equipment at all, although this isn't necessarily the easiest way to do it.

I don't feel at home working anywhere unless I have a few essentials to hand. I suggest you make sure the following items are nearby each time you start a study session:

✔ **A notebook:** When I was a student, a visiting academic gave me this advice: 'Get a notebook just for maths. Make notes on everything you do – the mistakes you make, the new things you learn, the results you get. You'll never regret it.'

Having a dedicated maths notebook makes picking up where you left off and remembering what you've worked on easier. Check out the 'Keeping Good Notes' section later in this chapter for more information on what to stick in your notebook.

✔ **Pencils, pens and felt-tips:** Having a notebook isn't much use without something to write in it with. I like to work in pencil if I'm drawing pictures or doing something I'm not confident in. I move on to ink when I'm happy with what I'm doing. The felt-tips are mainly for making bright memorable notes and highlighting the important things I want to remember.

✔ **A calculator:** In a lot of this book I show you how to work things out without using a calculator – after all, trying to figure out sums by hand is always a good exercise. That said, calculators are extremely useful for checking your answers.

✔ **A waste-paper basket:** This isn't philosophy. It's not all worth saving.

Check out the 'More tools of the trade' sidebar in this chapter for a list of optional extras you may find helpful.

A space of your own

Nothing in the rules of maths says you have to study at a desk. Or at home. Or indoors. In fact, changing your scenery can help put you into a different frame of mind and provide the inspiration you need. I often pick up my notebook and decamp to my local coffee shop or park in search of a breakthrough.

More tools of the trade

In addition to the necessities I outline in the section 'Tools of the trade', I like to have a few additional items at hand for study sessions. You might find the following things useful:

✔ **An MP3 player:** I don't like to be interrupted when I work, so I generally put my noise-reducing headphones on and crank up my music as loud as I can stand. Classical music tends to work best for me (otherwise the lyrics distract me), but work with whatever helps you concentrate. That could be silence – for some people, any kind of noise is off-putting . . . in which case, a pair of earplugs is the way forward. I know some people who work in cafes with headphones and no music, because you don't get so many people trying to make small talk if you look like you're listening to something else.

✔ **A computer:** The internet is choc-a-block with distractions, entertainment and useless information. But the Internet is also really good for looking up information and for finding exercises and explanations. If I can't work something out, I tend to ask Wolfram Alpha or Google for help and jot down the answer in my notebook so I don't need to ask again.

✔ **A glass of water and a bowl of fruit:** Discomfort isn't normally a good thing and can be a serious distraction when you're trying to learn. Getting hungry and thirsty isn't a good plan when I'm trying to concentrate, so keeping some healthy snacks nearby to tide me over helps me stay on task.

✔ **A guitar:** I wish I could tell you a guitar was an essential part of studying maths . . . but it isn't. For me, a guitar is an essential part of studying because it gives me a chance to switch off for a few minutes and let my brain work something through.

Discovering where you work best

Try out several places to see where you work best. Does the local library have quiet tables you could work at? Can you spread out on a bed? This was always my favourite when I was a teenage brat, as I could claim I'd worked myself to sleep. Maybe the kitchen table is a good place for you, or you work best with a notebook on your lap, half-distracted by the TV.

Try studying in different locations and at different times of the day. Get creative about mixing it up, and see what feels best for you.

Making your workspace better for studying

After you figure out where your favourite workspace is, you can tweak it to make the area even more conducive to effective studying. Here are three ways to improve where you work:

- ✔ **Minimise distractions:** Maths is a lot easier when you have undisturbed focus and you know you don't have to stop for any reason. When I need to concentrate, I turn off my phone and email, crank up some music and completely shut out the outside world for as long as I need. Sometimes I set a timer for 30 minutes or so, and sometimes I leave my study time open-ended. I do live alone though, which makes a distraction-free work zone a lot easier to achieve.

  If you have family or housemates who constantly demand your attention, you may need to cajole, bribe or intimidate them into leaving you alone when you study. I find a 'CAUTION: FRACTIONS!' sign tends to do the trick. Alternatively, you could just ask them nicely.

- ✔ **Maximise comfort:** If you're not physically comfortable when you study, relaxing mentally is tough. If your brain says 'I don't want to sit in that horrible plastic chair – it makes my back hurt,' persuading yourself to start studying is hard. Do all you can to make your workspace welcoming and cosy – make sure you can sit up straight, stay at the right temperature and have a drink. I keep a picture of my girlfriend on my desk – soppy as this sounds, it makes the prospect of sitting down to work just a little bit more inviting.

- ✔ **Keep things tidy:** Your mileage may vary with this, but when my desk is a shambles – which it quite frequently is – the mess distracts me . . . as does the nagging suspicion somewhere in my mind that my mum's going to visit and tell me to tidy up. I make a point of decluttering my desk before I start anything that needs my full attention.

  Of course, you might prefer to work in a mess. But don't come crying to me when my mum shows up and calls your room a bombsite.

Staying Motivated

Whenever I begin a new project, I start off with the enthusiasm of a small puppy with a new toy. After a while though – sometimes minutes, sometimes months – I start to lose interest a bit. If I'm not careful, I start skipping days and before long the project dies a quiet death.

Fortunately, I have a few schemes up my sleeve for avoiding the death of projects. The cheap psychological tricks that I describe in the following sections either keep you in the saddle or help you get back on the horse after you fall off.

Remembering why you're studying

The human brain loves the word 'because'. You're a (largely) rational creature, and having a reason for why you do something makes you more likely to do it – for example, saving up cash without a goal is harder than saving up for a new bike.

Think about why you're studying. Maybe you want to open a beauty salon, but you need a numeracy qualification before you can get on the right course. Maybe you want to give a presentation at work and be sure that your graphs are right. Maybe you want to help your kids with their maths homework. Now take a big bit of paper, write 'I want to understand maths because I want to open a beauty salon/give a presentation/help the kids with their homework,' sign your statement and put the paper somewhere you can't help but see it. Whenever you struggle with your studying, look at your note and remind yourself why you're learning maths. Your brain doesn't really care what your reason is. If you try to jump the queue at the photocopier, you may say 'Do you mind if I jump in, because I need to copy the agenda for my meeting in 10 minutes?' But the person in front of you is just as likely to agree if you say 'Do you mind if I jump in, because I desperately need to make some copies?'

Using the 'calendar of crosses'

When American comedian Jerry Seinfeld was an up and coming comic, he faced a problem that many comedians have: not enough material. To put this problem straight, he decided he was going to write something every day without fail. He picked up a cheap calendar and every day he wrote something, he marked the date with a big red cross.

After a few days, he found the crosses were making a chain – and the last thing he wanted to do was break the chain. The calendar helped motivate him to keep on writing and was probably a big factor in his success in his field.

You can apply this idea to any habit you want to form – doing more exercise, going to bed on time, drinking five glasses of water a day . . . or doing a few minutes of maths. You soon get into the habit of doing a few minutes a day just to avoid breaking the chain, even when you don't feel like studying.

Buy a cheap calendar or find one online to print out. Stick the calendar next to your reason for doing maths, and draw a cross on the calendar every day you do some maths work. Don't break the chain!

Rewarding yourself

Life coaches talk about the difference between 'towards goals' and 'away-from goals'. This is the same thing as carrots and sticks – if you want a donkey to move forward, you can bribe him with carrots or threaten him with sticks. Some donkeys – and people – respond better to bribes than to threats.

The 'calendar of crosses' (which I describe in the previous section) is a mixture of carrots and sticks – you have the reward of seeing a nice long unbroken chain, but you also have the threat that if you don't do something today, you'll break that chain.

Today, I'm tricking myself into writing on my birthday, which in any reasonable world would be a national holiday. I need to get this chapter finished, but after I reach my prescribed word count, I'll reward myself with a trip to my favourite coffee shop for a large cappuccino and some carrot cake. I'm drooling at the thought – but I'm writing a lot faster than I would without a (literal and metaphorical) carrot.

Getting Your Head On Straight

I have seen normal intelligent adults suffer full-blown panic attacks when asked to solve a maths problem. I've seen people freeze completely, and I've seen people deny they know how to add two and two.

Being intimidated by maths isn't unusual. I have moments where I look at a topic I'm not familiar with and tense up. Panic isn't the best place to be if you want to learn something – so I've discovered how to get myself into the right frame of mind for absorbing information and doing good work. In this section I describe a few of the tricks I use to sort myself out.

I like to think of the first three tactics as POPS: Posture, Oxygen and Positive Self-talk. Whenever you panic, try using POPS to calm you down.

Sitting up straight

You know when your teachers yelled at you for slouching? They may have had a point. Try the following exercise one day: go for a walk with your shoulders hunched and your head down, a frown of concentration on your face, hands deep in your pockets. Then walk with your shoulders back, head up, big smile, arms swinging freely. Feels better, doesn't it? There's a reason people tell you to smile when you pick up the phone.

Your posture plays a big part in how you approach a task and how well you do in that task. Sitting up straight, throwing back your shoulders and working with a smile all make it easier for you to work quickly and effectively. Try it and see!

Getting a breath of fresh air

I used to suffer from panic attacks – they're horrible and I wouldn't wish them on anyone. One of the tricks I was taught for dealing with panic attacks involved diaphragmatic breathing.

What breathing?

Yeah, I asked that too. *Diaphragmatic* means pertaining to the diaphragm, a muscle somewhere in your lower chest. The idea is that instead of taking shallow breaths into the top of your lungs, you take deep breaths as far down as possible and then breathe out slowly. Singers use diaphragmatic breathing before they go on stage: it has the twin benefits of helping lung capacity so they can sing better, and calming them down so stage-fright doesn't hit so hard.

Here's how to do it:

1. **Put one hand on the top of your chest and the other hand on your belly.**

2. **Breathe in as deeply as you can, trying not to move your upper hand. Your lower hand should move out as your lungs fill with air.**

3. **Breathe in for a count of seven, but don't hold your breath.**

4. **Breathe out very slowly for a count of 11.**

   Don't worry if you can't manage all the way up to 11 – just breathe out for as long as you can and make a note to breathe more slowly next time.

5. **Keep doing this for a minute or so and you'll feel your heart rate drop and your head start to clear.**

   A clear head makes maths a lot easier.

Talking to yourself – not as crazy as it sounds

One of my most effective tactics as a teacher is to convince students not to say 'I can't do this' but to say 'I can't do this *yet*' or 'I'm still learning this.' Describing yourself as being on the way somewhere really helps you carry on moving.

The way you express your thoughts can revolutionise how you work. People who go into a Monday thinking Mondays are rubbish tend to have rubbish Mondays. Sports people who think about losing win less often than those who think about winning. People who say 'I can do this when I've got the pieces in place' do better at maths than those who say 'I'm useless at maths. I'm stupid.'

You're not stupid. You've picked up a *For Dummies* book, which makes you a very smart person.

Learning from your mistakes

I was rubbish at learning to drive. I was sure I couldn't drive and used the mistakes I made in every lesson as evidence that I wasn't cut out for the whole car malarkey.

The problem was not that I was rubbish at driving but that I was beating down my confidence by focusing on my mistakes.

About the same time, I was learning to play guitar. I adopted a 'Right! I'll show you!' attitude after my parents told me I wasn't dextrous enough to play guitar and my music teacher decided I'd never amount to much.

In that mindset, every chord I hit that sounded good was a small victory, never mind the 12 bum notes that came before it. I was willing to try new ideas just to see what happened. I lapped up anything that could possibly make me a better guitarist.

A simple change of thinking can make all the difference – by accepting you'll make mistakes on the way to mastery and determining to learn from those mistakes, you may find your confidence grows by leaps and bounds.

And to the cyclist I, uh, encountered on my driving test? Apologies. I'll leave more room next time.

Keeping Good Notes

When I start working with a student, I can usually tell how interested they are in maths by taking a look at their notebook or exercise book. This isn't a foolproof rule (some of my brightest students have the same kind of messy scrawl as I do), but the neater – and more attractive – you make your notes, the easier you can read them, the easier you can see what you've done and the easier you can pick up on mistakes.

Deciding on a notebook or a computer

Where you keep your notes is up to you. Have a think about the following and see what sounds best for your way of working:

- **Notebook:** The advantage of working on paper is that your only limit is what you can do with pens and pencils. If you need a picture of a duck, you draw a duck. If you need to do some long division, you write out the sum. Maths isn't like English: the notes you take are completely different. Maths thrives on pictures and sums more than on sentences – although there is always some explaining to do. A notebook is pretty easy to carry around with you too.

- **Loose paper:** Working on loose sheets of paper can avoid the two negative issues associated with working in a notebook: organising your notes if you revisit a topic is difficult in a notebook, and 'I forgot my notebook' is a bit too easy to use as an excuse for not doing any work. Working on loose sheets of paper still has a drawback: losing sheets of paper or finding notes all over the house happens quite a bit. If you're much better at staying neat and organised than I am, using loose paper can be a really good approach.

- **Computer:** Taking notes on a computer is a bit trickier than using paper because drawing pictures and writing formulas on screen isn't easy. But working on a computer has advantages: if you have bad handwriting (like I do), typed notes are much easier to read – and you can search and edit your work without too many headaches.

I use a mixture of these approaches. When I learn something new, I take notes on paper or in a notebook. When I revise a topic, I type up notes on my computer.

Recording the language of maths

In some ways, maths is like a foreign language. Dutch, maybe. It has a lot of words in common with English, and some of it even looks enough like English for you to read it . . . but then there are places where it descends into unpronounceable chaos that makes no sense at all. Worse yet, some of the words you think you recognise aren't quite as they seem or have more precise meanings than you think.

Unfortunately, I don't know of anywhere you can go to immerse yourself in maths culture and language for a few months until you speak the lingo fluently. And there're only so many times you can watch *Good Will Hunting* before you go crazy.

Instead, I suggest you keep a vocabulary book in your notebook or on your computer. Whenever you come across a maths word that you don't know or don't quite understand, write down the word on one side of the page and write its definition on the other side.

Record cards are an excellent way to learn maths vocab. Write the word on one side of a card and the meaning on the other. Run through the cards every so often until you can rattle off the definition without thinking.

Acing the Exam

Just knowing your stuff isn't enough to do well in an exam. Being knowledgeable is just one leg of the trousers . . . and you don't want to go into an exam half-dressed.

You need to prepare for the exam as well. Think: what are they likely to ask? How long will I have? Will I be rushing, or will I have plenty of time to check? How will I keep calm and focused in the exam situation?

Nobody likes nasty surprises. The more practice you get with past papers, the better prepared you are likely to be. Most exam boards make past papers available for a small fee, and many offer them free of charge online. If you attend a maths class, ask your tutor to give you some papers to work through.

You may remember your schoolteachers lecturing you about the importance of exam technique. But, if your school was anything like mine, your teachers may not have mentioned much about what exam technique actually is. I may put myself at risk of the Teacher Secret Police coming after me by doing this, but here I reveal the highly classified details of exam technique:

✔ **Control your nerves:** Freezing up in an exam is not, generally, a good thing. I've been there. Luckily, I figured out how to unfreeze myself: I used the same techniques that I describe in the 'Getting Your Head on Straight' section: sit up straight, take a deep breath and tell yourself 'I can do it'.

POPS – posture, oxygen and positive self-talk – are fantastic tools for when you need to calm down quickly. Take a few moments to breathe and tell yourself 'I can do this' if you find yourself blanking in an exam – or anywhere else for that matter.

✔ **Read the paper:** If you prepare well, you'll know roughly what the test entails – but reassuring yourself that this test isn't different from the ones before is always a good idea. Reading through the paper first also gives your brain a chance to start its wheels turning on some of the harder questions – your subconscious can begin work on those while you rattle off the easy ones . . . which you can also pick out while you read through the paper.

✔ **Pick the low-hanging fruit first:** If you have any doubt about finishing on time, *do the quick, easy questions first.* You don't want to get to the end of the exam and realise the last two questions were straightforward and you could have answered them if you hadn't puzzled over a hard one you ended up guessing. Doing some easy questions first also gives you a good solid base – if you can pick up easy marks early on, your confidence grows and you have a little longer for the harder questions elsewhere.

✔ **Don't dilly-dally:** Don't spend too long on one question. If you get bogged down or stuck, leave a big star by the question, go on to the next question, and come back to the hard one later on if you have time. Spend time on marks you can definitely get. This also gives your mind some time to figure things out while you think about the next question – quite often you'll come back to the hard question and say 'Oh! It's easy!'

✔ **Sprint to the finish:** If you start to run out of time, stay calm and figure out how to use your last few minutes well. This advice applies especially in a multiple-choice test, where you can pick up a few extra marks by guessing wildly.

That's right – pick an answer. Pick all the same, pick all different or pick at random – it really doesn't matter. I talk about why in Chapter 19 on probability.

If you do a test that carries penalties for wrong answers however, guessing wildly may not be the best strategy. Even so, if you can cross out an impossible answer and pick from the others, you generally still do better by guessing than by giving no answer.

✔ **Check your answers:** With some multiple-choice questions, starting with the answers and seeing which one answers the question, rather than the other way round, is a good ploy.

Even if the exam isn't multiple choice, you can still check that an answer you reach by long division fits the question when you multiply it back. I show you how to do this double-checking for many of the topics in this book – look out for the TIP icons throughout the text.

✔ **Eliminate the impossible:** If you have five options, and four of them are clearly wrong, the fifth one must be right.

In practice, examiners aren't quite so generous as to offer only one obviously correct answer – but you can sometimes dismiss a couple of answers as obvious red herrings. For example, if the question involves cooking for more people, it's pretty likely that the new recipe calls for more of an ingredient.

Read the question carefully and make sure that what you write down answers the question.

Chapter 3

It All Adds Up: Addition and Subtraction

*A*dding things up and taking them away are the two most fundamental skills in arithmetic. If you master these skills – just two sides of the same coin – you'll find the rest of this book much, much easier than it would be without them.

Adding is what happens when you combine two groups of similar objects together. If I own four *For Dummies* books and I buy two more, I end up with six of them. 4 + 2 = 6. You add up when you gain or increase something.

One of the neat things about maths is that the rules hold whatever you add – it doesn't have to be *For Dummies* books. If you start with four cups of coffee and drink two more, you've drunk six cups of coffee. If you walk four miles and then walk two more, you've walked six miles. Whatever the things are, if you start with four of them and add two more, you end up with six.

You have to be careful when you're adding and taking away that the things you're working with are similar – you can't really add two apples to four oranges and get a meaningful answer without bending the rules (you could say it makes six pieces of fruit, but that's a bit of a cheat). It *really* doesn't make sense to add two clouds to four phones, or to add two kilometres to four grams.

You probably have a good idea about taking away too. Taking away, or *subtracting*, happens when you decrease, lose or spend things. If I have six *For Dummies* books and my cheapskate friend borrows two of them, I wind up with four books: 6 – 2 = 4.

In this chapter, I look at how to add and take away using a number line for small values. I also give you some hints on memorising some important sums. After all, counting on your fingers is totally acceptable – but it's much slower than being able to recall facts straight away. Imagine having to use a dictionary every time you wanted to spell any word at all . . .

In this chapter you get some practice at dealing with big numbers. I take some time to show you not only *how* to add and subtract big numbers but also *why* the methods work.

And last up, we look at negative numbers. I'm not going to make you do any weird sums, but in this chapter I help you understand what negative numbers are and how they work.

Nailing Down the Number Line

The number line is a very simple tool for adding and taking away small(ish) numbers. I give an example of a number line in Figure 3-1. The number line really is just a line with numbers written above it. Hard to imagine how it got its name! If you can count, you can draw a number line; if you can draw a number line, you can add and take away.

Figure 3-1:
The number line from 0 to 20. Numbers go on forever, so the line doesn't stop here – but you probably don't need to go any higher.

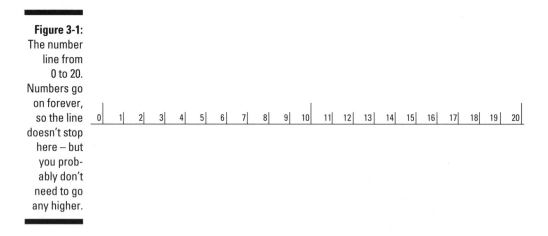

Don't rush – it's surprisingly easy to mess up the number line. If I try to use the number line in a hurry, I often find that I skip a number, which is terribly embarrassing if I'm trying to convince someone I'm good at maths. Look back at what you've drawn and count along the line to check you haven't missed or repeated a number.

Some people like to draw the number line vertically, starting with zero at the bottom of the page and counting up the line, like counting floors in a skyscraper. In this book, I work from left to right, mainly because it takes up less space.

I like to remember which way round the number line goes by saying 'the left is less (or lower)' – the three Ls.

You don't actually need to draw a number line. A ruler serves as a perfectly good substitute. A 30-centimetre ruler has all the numbers from 0 to 30 written out ready for you to count on as you please.

Adding and taking away with the number line

Adding and taking away with the number line are very similar processes. You just need to remember that *add* (+) means 'move to the right by' and *take away* or *subtract* (–) means 'move to the left'.

After that, you just count.

The number line is a first step to get you used to how adding and subtracting work. After you read this chapter, you may know your adding and subtracting number facts by heart – but safe in the knowledge that the number line's there if you need it.

Moving to the right and moving to the left

Maths is a very concise language – you can write down some incredibly complex ideas with just a few symbols. A lot of what you do as you learn maths is to translate maths into English – or into instructions for what you need to do.

Imagine you have eight pens in your pencil case. After you clean out your drawers, you realise you have another five pens. How many pens do you now have? You could make a big pile of pens and count them up – easy enough with pens, but not quite so easy with, say, wildebeest or space rockets.

Instead, you translate the problem into a sum: 8 + 5 = ? And then you translate the sum into instructions.

You translate this particular sum as 'start at eight and then move to the right by five.' You always start with 'start at', and the '+' sign translates as 'move to the right by'. The numbers just stay as numbers. Figure 3-2 shows what happens if you do this sum on the number line.

Figure 3-2: Doing 8 + 5 on the number line. Start by pointing at eight and count five steps to the right. You end up at 13, which is your answer.

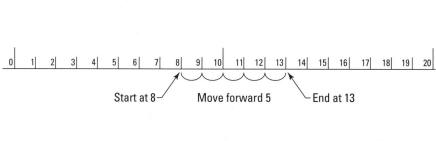

You see, you end up at 13 – which is your answer. Translating back to the original question, you wind up with 13 pens (or wildebeest or space rockets).

Here, in full, is the method for adding:

1. **Put your pen on the number line on the first number in your sum.**

2. **Move to the right by the number of spaces indicated by the second number in your sum.**

3. **Where you end up on the number line is the answer.**

Taking away is very similar – the only difference is that you translate the '–' symbol as 'move to the left by'.

Imagine I have £20 and spend £6 on coffee and a cake. I can use the number line to work out how much money I have left.

I translate the problem as 20 – 6 = ? and translate that as 'start at 20 and move to the left by 6.' Figure 3-3 shows you how.

Figure 3-3:
Doing
20 – 6 on
the number
line. Start
by pointing
at 20 and
count six
steps to the
left. You
end up
at 14, which
is your
answer.

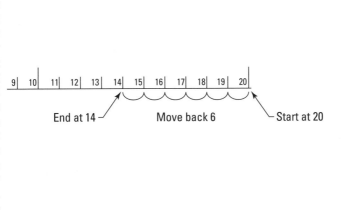

Using the number line gives me an answer of 14, which seems about right. I end up with £14 in change.

Here, in full, are the steps for taking away:

1. **Put your pen on the number line on the first number in your sum.**

2. **Move to the left by the number of spaces indicated by the second number in your sum.**

3. **Where you end up on the number line is the answer.**

The steps for adding and taking away are very similar – the only difference is the direction in which you move.

Seeing how close numbers are

Using the number line for taking away lets you see how far apart two numbers are. Some people call taking away 'finding the difference' – this method shows you why.

If you need to figure out 19 – 17, you don't really want to have to count back 17 spaces – you'll be there all night! Instead, look at the number line and see directly that you'd need to take two steps to get from 17 to 19 – so 19 – 17 = 2. Figure 3-4 shows you how this works.

This technique works only with taking away. Don't be tempted to try it with adding.

Figure 3-4:
19 – 17
using the
'difference'
method. You
mark 17 and
19 on the
number line
and notice
they're
two steps
apart, so
19 – 17 = 2.

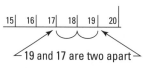

Adding and subtracting with two rulers

You can, with a bit of work, use two rulers to add and take away small numbers. To add two numbers (let's say 7 + 5, as in Figure 3-5), here's what you do:

1. **Find the first number (for our sum, seven) on one ruler.**

2. **Find the second number (for our sum, five) on the other ruler.**

3. **Put the two numbers next to each other.**

4. **Find the zero on either ruler and read the number it's next to (for our sum, 12). So, 7 + 5 = 12.**

Some rulers have centimetres on one side and inches on the other – make sure you use the centimetre side of both rulers!

Figure 3-5:
Adding
seven and
five with the
help of two
rulers.

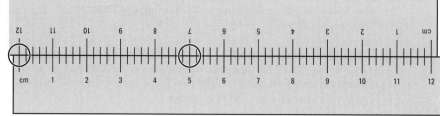

Taking away two numbers is slightly different. Let's do 7 – 5, as in Figure 3-6:

1. **Find the first number (for our sum, seven) on one ruler.**

2. **Put the zero of the other ruler next to the first number (for our sum, seven).**

3. **Find the second number (for our sum, five) on the second ruler.**

4. **Read off the number the second number is next to (for our sum, two). So, 7 – 5 = 2.**

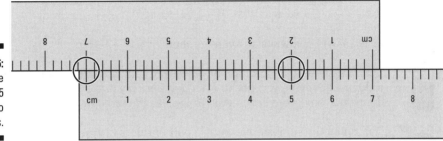

Figure 3-6:
Doing the
sum 7 – 5
with two
rulers.

Understanding Odd and Even Numbers

When I planned this chapter, I thought there could be hardly any real-life use for knowing about even and odd numbers. Then, this morning, I went out to try to find a new student's house. Wandering down Acacia Road, I noticed a sign saying 'ODD NUMBERS 15–37' – so I knew Eric's house (number 29) was in that particular side street.

You might also want to know about odd and even numbers if you play roulette.

An *odd number* is any number that ends in 1, 3, 5, 7 or 9. The *even numbers* are the numbers that end in 2, 4, 6, 8 or 0. So, 29 is an odd number and 0 is an even number. 1,000,000 is even. 1,000,003 is odd. You just have to look at the last digit.

Odd and even numbers have more use than simply finding Eric's house and breaking the bank at Monte Carlo. You can also use the concept of odd and even numbers to check whether your answer to an addition or take-away sum makes sense.

Try adding two numbers of the same 'flavour' – an odd number plus an odd number, or an even number plus an even number. Here are a few examples:

$$7 + 5 = 12$$
$$14 + 6 = 20$$
$$9 + 9 = 18$$

The answer is always an even number.

The same happens if you take away an odd number from an odd number, or an even number from an even number:

$$7 - 5 = 2$$
$$14 - 6 = 8$$
$$9 - 9 = 0$$

The answer is always an even number.

Any time you do an addition or take-away sum where the numbers are both odd or both even, you get an even number. Try using the following saying to help you remember: '*even* ground is all at the *same* height.'

Try doing the sums above using a pair of rulers or the number line if you want to practise the maths you know already. Now see what happens when we work with numbers of mixed flavours:

$$4 + 7 = 11$$
$$16 - 7 = 9$$
$$12 + 3 = 15$$

You get odd numbers. Perhaps you saw that coming. Whenever you add or take away numbers of different flavours, you get an odd number. Try using the following saying to help you remember this: 'the *odd* one out is *different*.'

Remembering Some Simple Sums

Using the number line or the ruler method that I describe earlier in this chapter is all well and good if you have all the time in the world. But when your sums involve big numbers, you may need to do several small-number sums one after the other. Before you know it, you've spent hours on a sum when you could have been taking a walk in the park.

I don't usually recommend remembering things you can easily look up, but you can save yourself literally days of work if you know your *number facts* – that is, all of the adding sums up to 10 + 10.

If you think this sounds like a lot of work, I'm afraid you're right: there are 100 of those pesky sums. I have some good news though: addition is *symmetrical* – it doesn't matter which way round you add things. For example, 4 + 7 is the same as 7 + 4 (they both make 11). So at least you only have to remember about half as many facts as you thought.

Meeting the adding table

+	0	1	2	3	4	5	6	7	8	9	10
0	0	1	2	3	4	5	6	7	8	9	10
1	1	2	3	4	5	6	7	8	9	10	11
2	2	3	4	5	6	7	8	9	10	11	12
3	3	4	5	6	7	8	9	10	11	12	13
4	4	5	6	7	8	9	10	11	12	13	14
5	5	6	7	8	9	10	11	12	13	14	15
6	6	7	8	9	10	11	12	13	14	15	16
7	7	8	9	10	11	12	13	14	15	16	17
8	8	9	10	11	12	13	14	15	16	17	18
9	9	10	11	12	13	14	15	16	17	18	19
10	10	11	12	13	14	15	16	17	18	19	20

Figure 3-7:
The adding
table for
numbers
1–10.

Figure 3-7 shows an adding table to help you learn your number facts. Adding with the table is easy – you find the first number you want to add in the top row and the second number you want to add in the left-hand column, and then follow down and across with your fingers until they meet.

You can take away with the table too, but this takes a little more practice: find the first number in the body of the table, and notice how that number is repeated through the grid in a diagonal stripe. Go along the stripe until the column you're in has the second number at the top of it. Read across to the left – the answer is at the start of the row. I show you how to solve 16 – 8 using this method in Figure 3-8.

Unfortunately, you won't always have the adding table with you. So you really do need to learn all the sums in it.

REMEMBER

You might think that spending time on adding and taking away small numbers seems a bit childish. Don't beat yourself up about having to learn stuff that seems basic. If you ever watch a football team training, you'll see the players spend hours making simple passes to each other and running back and forth. Most of the players have done this stuff since they were six, but they still practise so that passing and running is natural and easy when they play in a real match. Spending time reminding yourself how to do the basics of maths is absolutely fine. In fact I wish all my students would spend some time on this!

+	0	1	2	3	4	5	6	7	8	9	10
0	0	1	2	3	4	5	6	7	8	9	10
1	1	2	3	4	5	6	7	8	9	10	11
2	2	3	4	5	6	7	8	9	10	11	12
3	3	4	5	6	7	8	9	10	11	12	13
4	4	5	6	7	8	9	10	11	12	13	14
5	5	6	7	8	9	10	11	12	13	14	15
6	6	7	8	9	10	11	12	13	14	15	16
7	7	8	9	10	11	12	13	14	15	16	17
8	8	9	10	11	12	13	14	15	16	17	18
9	9	10	11	12	13	14	15	16	17	18	19
10	10	11	12	13	14	15	16	17	18	19	20

Figure 3-8:
Solving
16 – 8 using
the adding
table.

Learning your adding facts

Luckily, you can break down the adding table a bit. Start with the first column – adding one. You already know how to do that without thinking: you just go to the next number. So you hardly need to learn that column of the table.

Now look at the last row – adding ten. Adding ten is simple: you just put a one in front of the original number. For example, 2 + 10 = 12. And 7 + 10 = 17. (The exception is 10 + 10 = 20, but you probably know that.)

If you're happy to count backwards, you can add nine – just add ten and count back by one. So, to do 6 + 9, you can say '6 + 10 is 16. Count back 1 to get 15.' You can check this idea works with some of the other sums in the adding table.

So, now you actually only have 28 number facts to learn – but you may wonder how you will learn them.

I suspect that if you enjoyed sitting in rows at school reciting 'seven plus one equals eight, seven plus two equals nine . . .' for hours on end, you probably aren't reading this chapter. That method is effective . . . but deadly dull.

My way of teaching makes things a bit more interesting: I want you to play some blackjack:

1. **Find a pack of cards.**

2. **Deal two cards.**

3. **Find the score of each of the cards. An ace is worth one, and the court cards (jack, queen and king) are all worth ten.**

4. **Add up the scores of the two cards. If you can work out the score in your head without looking at the adding table or counting, pat yourself on the back. If not, write the sum in your notebook to look back at later.**

5. **Go back to Step 2 until you run out of cards.**

Play this card game every day for a week or so. The more you do it, the fewer questions you will need to write down in your book – which means you're learning. Go you!

Tackling your take-away facts

To learn the take-away facts, I need you to play some more blackjack – with a twist (see what I did there?):

1. **Find a pack of cards.**

2. **Deal three cards – two of them together and one below them.**

3. **Figure out the score of each of the cards. An ace is worth one and the court cards are worth ten each.**

4. **Add up the scores on the first two cards.**

5. **If the third card is smaller than your two-card total above, take away the third card score from the two-card score. If the third card is bigger than your two-card total above, take away the two-card score from the third card score.** For example, if you deal 6, 7 and 4, your sums are 6 +7 = 13, and then 13 – 4 = 9. If you deal 2, 3 and a king, your sums are 2 + 3 = 5, and then 10 – 5 = 5.

6. **If you work out the score in your head without looking at the adding table or counting, pat yourself on the back. If not, write down the sum in your notebook.**

7. **Go back to Step 2 until you run out of cards. (You'll have one left over, so just ignore that one.)** Do this subtraction game every day for a week – you'll quickly get the hang of taking away.

Increasing your success with flash cards

I like to use *flash cards* – little cards with a question or sum on one side and the answer on the back. You can buy ready-made flash cards, find free printable ones online, or make your own out of index cards or cut-up paper.

Here's how to use your flash cards to help your adding and taking away sums:

1. **Shuffle all the flash cards you want to practise.**

2. **Answer the question on the top one.**

3. **Check the answer. If you get it right straight away, put it to one side; if not, put the card to the back of the pile.**

4. **Go back to Step 2 until you've put all the cards to one side.**

The neat thing about the flash-card method is that it makes you practise the sums you need to practise more than the ones you know by heart.

Try working with flash cards against the clock. Seeing how quickly you get through the pack as you practise can be a great motivator.

What to do when you forget

Even with all the preparation in the world, you sometimes forget your sums. It happens to me, it'll happen to you, it happens to Stephen Hawking. Instead of getting upset or frustrated, finding other ways to approach a question is a good idea.

The best strategy is probably to go back to the two-rulers method or simply count on your fingers. But although these are perfectly good once-in-a-while strategies, you probably don't want to be doing them every time, otherwise your sums take hours.

When you forget a sum, write it down in your notebook. Simply writing down the sum and flagging it up as something you'd like to know better helps your brain get hold of the information.

Don't beat yourself up about forgetting or making mistakes! Very few people are lucky enough to learn everything perfectly the first time, or even the tenth.

Super-Size Me: Working with Bigger Numbers

The sums up to 10 + 10 are so important. Most of the sums you do involve small numbers (checking you haven't lost a finger in a washing-up accident, figuring out whether you have enough eggs for an omelette – the sums you do without even thinking about them). Importantly, these small-number sums are key building blocks to working with bigger numbers.

Adding and taking away bigger numbers

For the purposes of this section, imagine travelling in a country where only three types of banknote are available: £100, £10 and £1.

In your cash drawer you have £125 – a £100 note, two £10 notes and five £1 notes. You can read the first number of £125 as how many £100s you have, the second number as how many £10s and the third as how many £1s. For your birthday a generous friend gives you £213 to celebrate. He gives you two £100 notes, one £10 note and three £1 notes.

You had one £100 and have just added two, so now you have 1 + 2 = 3; three £100 notes.

You can do the same for the £10s: you had two and your friend gave you one, so now you have 2 + 1 = 3; three £10 notes. Likewise you had five £1 notes. Your friend gave you three, so now you have 5 + 3 = 8; eight £1 notes.

Altogether, you have three £100s, three £10s and eight £1s, making a grand total of £338. Notice how you can just write down the numbers of each note to give you your final answer.

Taking away uses the same kind of idea. You have £975 to last the month and your landlord asks for rent of £702. In your cash drawer you find nine £100 notes. The landlord wants seven, so you have 9 – 7 = 2 left.

You have seven £10s. The landlord doesn't want any of them, so you have 7 – 0 = 7 left.

And you have five £1 notes, of which you have to give away two, so you have 5 – 2 = 3 left.

Altogether, that leaves you with £273.

Figure 3-9 shows how to write out the sums that I describe above.

Figure 3-9:
Examples of adding and taking away bigger numbers.

```
  1  2  5        9  7  5
+ 2  1  3      - 7  0  2
  3  3  8        2  7  3
```

Knowing your adding and taking-away facts up to 10 + 10 and 20 − 10 will be extremely useful in this section. If you're not confident, keep practising – it'll soon come!

Following recipes for adding and subtracting

I have horrible, spidery, scruffy handwriting that means almost nobody can read my rough work. I single-handedly got my school rules about using fountain pens changed because all I could hand in was a mess of smudged inky pages, and my hands turned everything I touched blue. I take a sort of perverse pride in my scruffiness.

The only thing I deliberately do neatly is laying out big sums. I know that if I get my columns mixed up – my tens mixed up with my hundreds, say – I go disastrously wrong.

For the sake of your sums, try to keep them neat. Give yourself plenty of space so you don't mix up columns. Use squared paper if it helps.

Adding larger numbers

The basic recipe for adding numbers is to split them up into hundreds, tens and units – or £100 notes, £10 notes and £1 notes – and add each type (or each kind of banknote) separately.

This is great – as long as you never have more than ten of each type of banknote.

To deal with this happy situation, you have to *carry* – or, if you prefer, 'make change'.

Imagine you have £9 and someone gives you £3. You know that gives you £12. Perhaps you have 12 £1 notes, or you may have a £10 note and two £1 notes. If you have more than ten of a banknote, you can change ten of them into one banknote of the next value up. Ten £10 notes are the same as one £100 note. Ten £1,000 notes would be the same as one £10,000 note.

The way we deal with this in a sum is to add a one below the next column to the left, as in Figure 3-10.

Figure 3-10:
Adding with
a carry.
Instead of
writing 15
at the
bottom
of the
right-hand
column, you
carry the 1
and write
it below
the next
column. You
then add
the 1 in
that next
column –
here, that's
6 + 2 + 1 = 9.

```
    6   6              6   6
 +  2   9           +  2   9
        5              9   5
 ........................................
    1                  1
```

Here's a recipe for adding big numbers:

1. **Line up the two numbers you want to add so their ends (the *units* or £1 notes) are in line, just like in Figure 3-10.**

2. **Add up the right-most column.**

3. **If your total is less than ten, write the total under the column. If the total is ten or more, write the second digit under the column and put a 1 under the next column to the left.**

4. **Go to the next column and add up the numbers – and add 1 if you've written '1' below this column.**

5. **Repeat Steps 3–4 until you run out of numbers.**

Subtracting larger numbers

A similar problem springs up when you try to take away. Say you need to work out 26 – 7. You start by trying to work out 6 – 7, but you can't do it: if you have six £1 notes, you can't give seven of them away – you don't have enough.

Instead, you take one of your two £10 notes and change it for ten £1s, so you have one £10 note and 16 £1s. You can now take away seven £1s to leave nine £1 notes and the one £10 you didn't touch – making 19.

Keeping track of all of that when you write out a sum is pretty hard – which is one of the reasons you need to give yourself plenty of space when you do maths. To show a carry in your take-away sum, here's what you do:

1. **Write a small 1 above the number you want to take away from. In the example in Figure 3-11, you end up with a 16. The 1 represents the ten you've borrowed from the next column.**

2. **Cross out the number to the left, reduce it by one and write the new number above and to the right.**

3. **Carry on as before.**

Figure 3-11:
A take-away sum with a borrow. You can't do 6 – 7, so you borrow a ten from the next column to the left.

$$
\begin{array}{cc}
2 & 6 \\
- & 7 \\
\end{array}
\qquad
\begin{array}{cc}
\cancel{2}^{1} & {}^{1}6 \\
- & 7 \\
\hline
1 & 9 \\
\end{array}
$$

Borrowing gets tricky when you try to borrow from a number that's not there – say, in the sum 503 – 6. You want to change a £10 note into £1s so you can do the final column – but there aren't any £10s.

The solution is to change one of the £100 notes into ten £10s, and then change one of those £10s into ten £1s – as in Figure 3-12.

Here's my recipe for taking away numbers:

1. **Line up the two numbers you want to take away so their ends (the units or £1 notes) are in line.**

2. **Take away the right-most column. If necessary, borrow a ten from the column to the left and then do the take-away sum.**

3. **Write your answer under the column.**

4. **Go to the next column and repeat Steps 2–3.**

5. **Repeat Steps 2–3 until you run out of numbers.**

Figure 3-12: A take-away with a double borrow. You can't do 3 – 6, and you can't borrow from the tens because there aren't any. You create some tens by borrowing from the hundreds, and then borrow from one of those newly created tens.

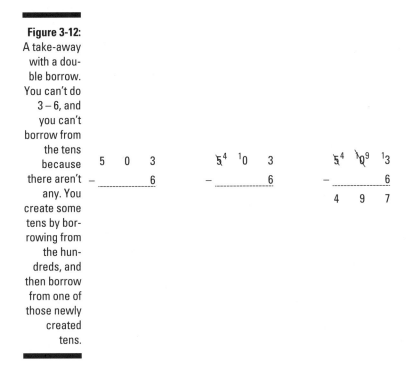

Going Backwards with Negative Numbers

In most of this chapter – and most of this book – I talk about *positive numbers*, or numbers bigger than zero. But things also happen below zero.

Negative numbers (sometimes called minus numbers) are numbers that are below zero. For example, you get the number –3 ('negative three' or 'minus three') when you take three away from nothing: a temperature of –3 degrees is three degrees below zero.

Take a look at the number line that I describe earlier in this chapter. You see that I have chopped a bit off it. The number line actually carries on forever in both directions – to the right, the numbers get bigger; to the left, the numbers become negative, as in Figure 3-13.

The interesting – by which I mean slightly difficult – thing with negative numbers is the bigger the number looks, the smaller it actually is. If you go outside in –10 degree weather, you get colder than if you go out in –5 degree weather, even though the number ten looks bigger than the number five.

The negative sign turns things around – so what looks bigger is actually smaller. Negative numbers are always smaller than positive numbers.

You won't need to do too much maths with negative numbers in day-to-day life, but here are some contexts where you may meet them:

- **Temperatures:** Temperatures are by far the most common reason you use negative numbers in day-to-day life. At zero degrees Celsius, water freezes and things get frosty. As the temperature gets colder, the frost gets worse and the number gets bigger.

  Last night, where I live, the temperature dropped to –3 degrees. But that's not as cold as things got when I lived in Montana – one night there, the temperature dropped to –35 degrees, so cold I had to install a heater in my car engine to stop it freezing solid. A typical freezer has a temperature of about –20 degrees, which is warmer than the Montana winter but cooler than the British one.

- **Goal differences:** You find negative numbers in football league tables. A team's goal difference is how many more goals it scores than it concedes. A team towards the bottom of the league usually concedes more goals than it scores, so the team has a negative goal difference – the bigger the negative number, the worse the team is doing, so –30 is a worse goal difference than –5, and both are worse than a goal difference of 1.

- **Money:** You may see 'less than no money' written as a negative number. When my bank balance drops below zero and I am overdrawn, my statement shows my balance as a negative number. The bigger the negative number, the worse my finances: a balance of –£500 (£500 overdrawn) is worse than a balance of –£50 (£50 pounds overdrawn), and both are worse than a balance of £1 (£1 in credit).

- **Changes:** When a number decreases over time, you may see it written as a negative change. For example, if house prices in a particular area go down by 2 per cent, you may see this written as '–2%' rather than '2 per cent down'. A price change of –10% is worse for the seller than a change of –5%, and both are worse than no change (0%) or a small increase (1%).

Chapter 4

Equal Piles: Multiplying and Dividing

*I*magine you're at the supermarket and buy four boxes of eggs, each containing 12 eggs. You need to work out how many eggs you have altogether.

The long way to do it is to start with 12 + 12 (making 24), then add another 12 (making 36), and then add a fourth 12 – giving a total of 48.

Working like that is okay if you have only a few boxes of eggs. But if you own the supermarket and get an order for 75 boxes of eggs, you may not want to spend your whole morning adding things up.

Instead, you can *multiply* or *times* 12 by 75 – which we write as 12×75. In this chapter I show you how to multiply small numbers and then how to use that skill to multiply bigger numbers.

Multiply and times mean exactly the same thing and you can use the terms interchangeably. 'Multiply' is more technical but 'times' is easier to say.

Now imagine you have 3,600 eggs and need to put them into boxes of 12. To work out how many boxes you need, you could simply count down 12 at a time until you get to zero – but that may take all day. Instead, you can divide 3,600 by 12. In this chapter I tell you all you need to know about dividing.

You may be pleased to know you won't find much algebra in *Basic Maths For Dummies*. I do need to show you how to use formulas though, so I cover formulas very quickly at the end of this chapter. Don't worry – I'll be gentle.

Meeting the Basics of Multiplication and Division

The prefix *multi-* means 'many' – think 'multi-player game' or 'multiplex cinema'. The suffix *-ply* means times – for example, two-ply tissues. *Multiply*, then, means 'many times'.

'Nine times seven' simply means 'add up a list of nine sevens' – or 'add up a list of seven nines'. You can see any multiplication sum as a number of equal piles of things: the first number is how many piles you have, and the second number is how many things you have in each pile.

You may have memories of reciting times tables at school. I was terrified of my maths teacher at secondary school, even though I was good at my times tables. Every Friday, he spent half the lesson marching up and down barking out 20 questions from the times tables. We learnt them soon enough, but teaching by intimidation is hardly the *For Dummies* way.

Instead, in this chapter I show you the times tables and – just like with the adding tables – give you some games to play to remember them. The times tables usually only go up to ten, so I also show you how to work with bigger numbers. Things are spookily similar to the adding I explain in Chapter 3.

Dividing is exactly the opposite of multiplying: you take a number of things and split them into equal piles. Armies are split up into divisions. So are football leagues.

'Ninety-two divided by four' just means 'split up 92 into 4 piles and tell me how big the piles are'. Or, 'split up 92 into piles of 4, and tell me how many piles there are'. In this chapter I show you some games to help you remember your division sums up to $100 \div 10$, and then show you how to do division when you have bigger numbers. Again, you just need to split up piles.

Remembering Your Times Tables

In Chapter 3 I show you how to add big numbers by adding up a series of smaller adding sums. In this chapter I show you how to multiply big numbers by adding a series of multiplication sums. So, first of all, you need to know your *times tables* – all the times sums up to 10×10. You may think this sounds like a lot to remember – 100 sums! But you have to do a lot less than you think. The times table, which I show in Figure 4-1, is *symmetrical*, so 4×8 is the same as 8×4 (both sums make 32).

x	0	1	2	3	4	5	6	7	8	9	10
0	0	0	0	0	0	0	0	0	0	0	0
1	0	1	2	3	4	5	6	7	8	9	10
2	0	2	4	6	8	10	12	14	16	18	20
3	0	3	6	9	12	15	18	21	24	27	30
4	0	4	8	12	16	20	24	28	32	36	40
5	0	5	10	15	20	25	30	35	40	45	50
6	0	6	12	18	24	30	36	42	48	54	60
7	0	7	14	21	28	35	42	49	56	63	70
8	0	8	16	24	32	40	48	56	64	72	80
9	0	9	18	27	36	45	54	63	72	81	90
10	0	10	20	30	40	50	60	70	80	90	100

Figure 4-1:
The times
tables up to
ten.

Follow these steps to multiply with the times table:

1. **Find the first number you want to times in the top row.**

2. **Find the second number you want to times in the left-hand column.**

3. **Follow down and across with your fingers until they meet.**

4. **Where your fingers meet is your answer.**

If you've used the addition tables in Chapter 3, the times tables work just the same way.

You can divide with the times tables too – but this is a bit harder:

1. **Find the second number (the number you want to divide *by*) in the left-hand column.**

2. **Move to the right along the table until you find the first number in the sum, the one before the divide sign.**

3. **Go up from there and find the number at the top of the column – that's your answer.**

If the first number isn't there, the sum doesn't come out exactly – so if you try to split the number into that many piles, you have some left over. For example, if you try to split 13 sweets between three children, everyone gets 4 sweets (using up 12 of the sweets) and there's 1 sweet left over. In real life, you may eat that yourself to save on arguments. In maths, you might write 13 ÷ 4 = 3 r 1. The 'r' means 'remainder' – or 'what's left over'. So 3 r 1 means 'Everyone gets three, and there's one left over.'

To find the answer when a divide doesn't come out exactly, here's what you do:

1. **Find the second number (the number you want to divide by) in the left-hand column.**

2. **Find the biggest number in that row that's *smaller* than the number you're trying to find.**

3. **Take away the smaller number from the bigger. This gives you the remainder: *how much* smaller it is.**

4. **Write down 'r' and then whatever the remainder is.**

5. **The first bit happens just like before: you read up from your smaller number.**

6. **Write the number at the top of the column in front of the r.**

I give an example of this in Figure 4-2.

Figure 4-2:
Working out
70 ÷ 8 using
the times
tables. 64 is
the biggest
number
in the row
less than
70 – read up
and find the
8. It is 6 less
than 70, so
the answer
is 8 r 6 – or
eight with
remainder
six.

x	0	1	2	3	4	5	6	7	8	9	10
0	0	0	0	0	0	0	0	0	0	0	0
1	0	1	2	3	4	5	6	7	8	9	10
2	0	2	4	6	8	10	12	14	16	18	20
3	0	3	6	9	12	15	18	21	24	27	30
4	0	4	8	12	16	20	24	28	32	36	40
5	0	5	10	15	20	25	30	35	40	45	50
6	0	6	12	18	24	30	36	42	48	54	60
7	0	7	14	21	28	35	42	49	56	63	70
8	0	8	16	24	32	40	48	56	64	72	80
9	0	9	18	27	36	45	54	63	72	81	90
10	0	10	20	30	40	50	60	70	80	90	100

Practising your times tables

Ideally you need to remember all of the times tables. I know – it's an insufferable chore and a terrible amount of work and so unfair – at least, so my students tell me. Things aren't so bad though.

The one times table is just the numbers themselves. So, $1 \times 1 = 1$, and $1 \times 2 = 2$, and $1 \times 3 = 3$, and so on. The ten times table is similar, but with a zero stuck on the end – so $10 \times 1 = 10$, and $10 \times 2 = 20$, and $10 \times 3 = 30$.

For the other numbers, I need you to play some blackjack:

1. **Find a pack of cards.**

2. **Deal two cards and find the score for each – aces count as one, and jacks, queens and kings are all ten each.**

   To make the game harder but shorter, whenever you get an ace, jack, queen or king, deal another card on top straight away.

3. **Times the numbers of the two cards together.**

4. **If you get the answer quickly, smile broadly. If not, write down the sum in your notebook.**

5. **Go back to Step 2 until you run out of cards.**

Try playing this game once a day for a week or two and see how it gets easier the more you play.

Working around mind blanks

I'm supposed to be a hot-shot maths genius. Every so often, if I'm tired or distracted or overconfident, I write down something like $2 \times 3 = 5$ in class and my students laugh at me. It's okay to laugh at your teacher when they do that. The point is, even experts forget their times tables once in a while.

So don't panic if you have a mind blank every so often. The important thing is to be able to work out the sum numbers if you blank.

The most common way to figure out (say) your seven times table is to count up in sevens while counting on your fingers: 1 times 7 is 7, 2 times 7 is 14, 3 times 7 is 21, and so on. If you're confident with your adding (which I cover in Chapter 3), this is a reliable (but slow) way to do it. This method also has the advantage that by saying '5 times 7 is 35', you remind yourself of your times-table facts as you go along.

I also have a couple of nice ways to remember certain times tables using my fingers. Luckily, these tricks work for the more difficult times tables (towards the bottom and the right of the table).

Folding under your nines

You can figure out your nine times table by folding under your fingers as you go along. Here's how it works:

1. **Put both hands on the table in front of you, pointing away from you.**

2. **Mentally number your fingers from one to ten.**

   Your left pinky is one, your left thumb is five, your right thumb is six and your right pinky is ten.

3. **Ask the question 'What am I multiplying nine by?'**

   For example, for 9×7, you multiply nine by seven.

4. **Fold the finger with that number under your hand.**

   In our example, the number is seven, so you fold your right index finger. (I show what I mean in Figure 4-3.)

5. **Write down how many fingers and thumbs are to the left of your folded finger. This is how many tens are in the answer.**

   In our example, you have six fingers to the left – all of your left hand plus your right thumb.

6. **Write down how many fingers and thumbs are to the right of your folded finger. This is how many singles are in the answer.**

   In our example, you have three fingers to the right.

7. **You've just written down the answer. Awesome.**

   In our example, your answer is 63.

The finger-folding method works only with the nine times table. If you try it with other times tables, you get the wrong answer.

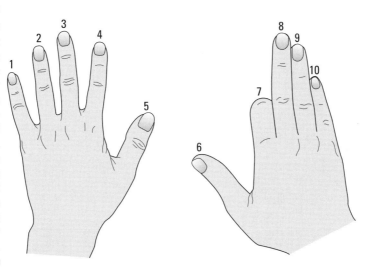

Figure 4-3:
Working out
7 × 9 with
the folding-
finger
method. The
seventh fin-
ger is folded
underneath.
There are
six fingers
to the left
and three
to the right:
the answer
is 63.

Pointing at the big ones

The neat little trick I describe here works for the sums from 6 × 6 up to 10 × 10. Here are the steps:

1. **Put your hands on the table, with your fingers pointing towards each other.**

2. **Mentally number the fingers on each hand.**

    Your thumbs are six and your little fingers ten.

3. **Touch the two fingers you want to multiply together to each other.**

4. **Count how many fingers on your left hand are above (on the pinky side) your touching fingers.**

5. **Count how many fingers on your right hand are above (on the pinky side) your touching fingers.**

6. **Times those two numbers together and write down the answer.**

7. **Count up the remaining fingers on both hands, including the touching ones.**

    This is how many tens you have, so write down that number with a zero after it.

8. **Bingo! Add the two numbers and you have your answer.**

If this sounds ridiculously complicated, have a look at Figure 4-4, where I show you an example for the sum 7×8.

Figure 4-4:
Working out
7×8 with
the touch-
ing-finger
method.
Above the
touching
fingers,
there are
two fingers
on one hand
and three
on the other;
these times
together to
make six.
Including
the touch-
ing fingers,
there are
five fingers
below, so
we add 50
and find the
answer to
be 56.

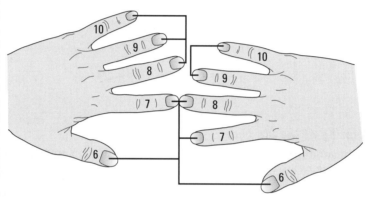

You count up the fingers below and including your touching fingers to get the tens and times together the numbers above to get the singles.

Working backwards

A useful strategy for working out a forgotten times sum is to work backwards from a sum that you do know. For example, if you know that $7 \times 6 = 42$, you know that 7×7 is simply one more 7 – making 49. You can also see that 8×6 is one more 6 – making 48.

Here's the recipe when you can't remember a sum in the times table but know the one above it or below it:

1. **Work out which number is common to both of the sums.**

    For example, if you use 5×8 to work out 4×8, 8 is the common number.

2. **Work out whether the number you want is higher or lower than the one you know.**

    If you want a higher number, you need to add something on; if you want a smaller number, you need to take away.

3. **Either add or take away the common number from the answer you know.**

Multiplying Bigger Numbers

Just like with adding (which I talk about in Chapter 3), learning your multiplication facts is a step towards working with bigger numbers. Think how frustrating it would be if you had to stop multiplying just because a sum went all the way up to 11.

In this section I describe two stages of difficulty: multiplying a big number by a smallish number (from one to ten), and multiplying two big numbers – which isn't difficult but is quite involved. And repetitive. And fiddly. It takes a while, which is why it's called *long multiplication*.

I give you a couple of ways of working with large times sums. Choose the one you find more useful and appealing.

Multiplying a big number by a small number

Most ways of multiplying numbers together work much the same way: you think of the bigger number as money and the smaller number as piles. You split the money into 'banknotes' (singles, tens, hundreds and so on) and figure out what you end up with when you pile up each banknote the right number of times.

Piling up the money

Say you want to give each of your four cousins £123 for Christmas. You need to work out how much money you need.

Start by splitting £123 into one £100 note, two £10 notes and three £1 notes. You have four cousins so you need four times as many of each note.

So, you need four £100 notes, eight £10 notes and 12 £1 notes. The £100s add up to £400, the £10s add up to £80 and the £1s add up to £12. When you add those up, you get £492.

Ignore the zeros until the end. When you want to work out 4×20, think of four piles, each with two £10 notes. That way, you just work out $4 \times 2 = 8$; so you have eight £10 notes, making £80.

Using the grid method

Another way of doing bigger times sums is to use the *grid method* – which happens to be my favourite. The grid method is also called the *lattice* method. Here's the recipe:

1. **Draw out a grid as many squares wide as the bigger number has digits, and as tall as the number of digits in the smaller number.**

   For example, for 123×4, the grid is three squares wide and one square tall. Give yourself plenty of room to work.

2. **Split the big number into banknotes as I describe in the previous section and write the amount of each at the top of each row.**

   For our example, you write 100 at the top of the first column, 20 at the top of the second column, and 3 at the top of the third column.

3. **Write the small number at the far left of the row.**

4. **In each square of the grid, write down what you get if you times the number at the top of the column by the number at the left-hand end of the row.**

   Remember to ignore the zeros until the end.

5. **Add up all of the amounts you have to see your answer.**

I show an example of the using the grid method in Figure 4-5.

Figure 4-5:
Doing
123×4
using the
grid method.

x	100	20	3
4			

x	100	20	3
4	400	80	12

```
400
 80
+12
___
492
```

Working the old-fashioned way

The old-fashioned way for multiplying a bigger number – the way I was taught at school – looks like this:

1. **Write down the big number.**

2. **Below the big number, write down the smaller number.**

   Make the ends of the two numbers match up, as in Figure 4-6.

3. **Working from right to left, times each digit in the big number by the small number.**

   Write down the answer so that the last digit lines up with the digit in the top number. The last digits will form a staircase pattern.

4. **Add up the numbers you've written in each column, carrying as necessary.**

   I explain carrying in detail in Chapter 3.

5. **There's your answer!**

Here's what it looks like:

Figure 4-6:
Doing
123×4 the
old-fash-
ioned way.
First do $3 \times$
4, then
2×4, then
1×4 –
notice the
staircase
pattern on
the right.
Finally, add
up each
column
to get the
answer,
492 – the
same as
before.

```
      1   2   3
  ×           4
          1   2
      8
  + 4
      4   9   2
```

Generating a times table

I can see that frown of confusion on your face after you read that heading. Why would *anyone* want to generate a times table? You can create a times table for multiplying and dividing big numbers. In the section above I show you to times repeatedly by the small number. If you need to multiply by a bigger number, you save yourself a bit of time if you work out the times tables before you begin and then simply look up the number you need to write down.

One way to do this is to work out the sums by multiplying in your head. Sit still for ten minutes doing 47×1, 47×2, 47×3 and so on until your brain melts. I don't really recommend this method, but you can do it if you need to.

The second way is a bit quicker and less brain-melty: write down the number you want to create the table for next to a number one in a circle. Then add the number to itself and write the answer next to a circled two. Add the original number on again and keep going until you get to a circled ten. You should have the original number with a zero on the end – if not, you've made a mistake somewhere and need to go back and check.

Multiplying two big numbers

Multiplying two big numbers is very similar to multiplying a big number by a small number, except one of the numbers is bigger than before.

Think about taking the bigger number, splitting it up into 'banknotes', work out how much money is in each of the piles, and add those up. I describe this idea in more detail in the section 'Piling up the money' earlier in this chapter.

Expanding the grid

Have a look at the earlier section 'Using the grid method'. You can apply this method to multiplying two big numbers. Here's what you do:

1. **Draw out a grid as many squares wide as the bigger number has digits, and as tall as the number of digits in the smaller number.**

   If you want to do 456×78, you need a three-by-two grid. Give yourself plenty of space.

2. **Change the big number into banknotes and write them across the top.**

   In our example, write 400, 50 and 6.

3. **Do the same with the smaller number, but at the left-hand end of the rows.**

   Here you have 70 and 8.

4. **In each square, times the first digit from the top of the column by the first digit at the start of the row and follow it by as many zeros as they have between them.**

   For 400×70, you do $4 \times 7 = 28$ and then follow it with three zeros (two from the 400 and one from the 70). Your answer for that square is 28,000.

5. **After you fill all the squares, add up the totals in the squares to reach your final answer.**

I show you an example of multiplying two large numbers using a grid in Figure 4-7.

x	400	50	6
70	28000	3500	420
8	3200	400	48

```
28000
 3200
 3500
  400
  420
+  48
35568
```

Figure 4-7: Long multiplication with a grid.

Going back in time

In the olden days, when shell suits were the height of fashion and *Blackadder* was still new, I learned to do long multiplication at school. Here's how it works:

1. **Write the bigger number above the smaller number and line up the last digits.**

2. **Work out the times table for the smaller number.**

3. **Times each digit of the bigger number by the smaller number by looking it up in the times table.**

4. **Write the answer underneath.**

   Line up the end of the new number with the digit you were multiplying by – you should get a staircase effect, as in the example in Figure 4-8.

5. **Add up the numbers that result to see your answer.**

Figure 4-8:
Long multiplication the old-fashioned way. Look up 6×78, 5×78 and 4×78 in the grid and write them below – notice the staircase pattern. Then add the columns up.

```
        4   5   6
    ×       7   8
        4   6   8
    3   9   0
3   1   2
3   5   5   6   8
    1
```

One for You, One for Me: Handling Division

You do a divide sum whenever you want to split a quantity into smaller, equal parts. Division is the exact opposite of multiplication. If $123 \times 4 = 492$ means 'if you make four piles of 123, you need 492', then $492 \div 4$ means 'if you split 492 into four piles, how big is each?' – the answer is 123.

Dividing and conquering

As with all of the sums you tackle in this book, division is much easier if you break it down into smaller pieces. And, just like with everything else in this book, I find the sums are easier if you think of numbers as representing money – £1 notes, £10 notes, £100 notes and so on.

Imagine you have £126 you want to split between seven people. You begin with one £100 note, two £10 notes and six £1 notes.

You try to divide the £100 note fairly between seven people, but you can't do this because you have only one £100 note. Instead, you turn the £100 note into ten £10 notes. With the two £10 notes you had before, you now have 12 £10 notes altogether.

Twelve is more than $1 \times 7 = 7$ but less than $2 \times 7 = 14$, so you can give everybody one £10 note. They all say thank you, and you have five £10 notes left over. You change the five £10 notes into 50 £1s, making £56 altogether because you also have six £1 notes. Fifty-six is 8×7, so everyone gets eight £1 notes.

Each person now has a £10 and eight £1s, making a total of £18, which is your answer.

Taking one step at a time

When you divide, you work from left to right, the complete opposite of what you do with all other sums in this book.

Here's the process for doing a division sum:

1. **Write the big number (the one you want to split up) under a 'bus stop' like the one in Figure 4-9.**

   Leave plenty of space between your numbers – you'll probably be writing more numbers in between them.

2. **Write the smaller number in front of the bus stop.**

3. **Somewhere nearby, write out the times table for the small number.**

4. **Find the first number under the bus stop that you haven't worked on yet.**

   Remember to work from left to right.

5. **Look for the biggest number in the times table that's smaller than your target, and the single-digit number beside it in the list.**

   Write the single-digit number above your target, on the roof of the bus shelter.

6. **Figure out how many are 'left over'.**

   Find the difference between the number you found in the times table and your target number. Write this just to the left of the next number under the bus stop to make a new number – for example, if you have six left over and the next number is three, the next number you work with is 63.

7. **Go back to Step 4 and repeat until you run out of numbers.**

8. **The number above the bus stop is your answer.**

$$7 \overline{)1 \quad 2 \quad 6}$$

$$\begin{array}{c} 0 \\ \hline 7 \overline{)1 \quad {}^1 2 \quad 6} \end{array}$$

Figure 4-9:
Division by a
single-digit
number.

$$\begin{array}{c} 0 \quad 1 \\ \hline 7 \overline{)1 \quad {}^1 2 \quad {}^5 6} \end{array}$$

$$\begin{array}{c} 0 \quad 1 \quad 8 \\ \hline 7 \overline{)1 \quad {}^1 2 \quad {}^5 6} \end{array}$$

As with all of your sums, keep your working clear and tidy so you can spot and correct any mistakes. Leave plenty of space between the numbers so you can fit in any extra digits you need.

Dealing with the left overs: Remainders

Imagine your baker has sold you a baker's dozen of doughnuts – that is, 13 – for your office party, where you plan to feed six people. Naturally, in the interests of fairness, everyone gets two doughnuts – but there's one cake left over.

Realistically, you can take two approaches: you can dispose of the doughnut and pretend it's not part of the sum (perhaps by eating it on the sly); or you can try to split up the doughnut into six equal pieces.

The first approach (leaving it alone) is one way of dealing with the problem of left-over numbers when you divide mathematically. You may say '$13 \div 6 = 2$, with one left over', or even '2, remainder 1'. Remainder just means 'what's left over'.

Depending on what question you're trying to answer, this is often a perfectly good answer. Sometimes, though, you need to do some more with the remainder and turn it into a fraction. Don't cry, they really don't bite, honestly.

Checking your answer

When you divide, get into the habit of checking your answer is right. The quick-and-easy but not-very-good-for-learning way is to redo the sum on a calculator.

Alternatively, if you want to practise a little more, try multiplying your answer by the small number and check you get back to the big number you started with.

In either case, you get a big burst of self-esteem every time you get a sum right. If you happen to get a sum not so right, don't worry – keep at it! Try the sum again, or try another sum and come back to the original sum later.

However, if proof were needed that professional mathematicians are not like normal people, pretty much everyone I've ever worked with would go for the 'chop the remainder up into equal slices' solution.

This is to say, they'd all turn the remainder into a fraction. This is honestly a really easy process:

1. **Working from left to right: write down the remainder.**
2. **Write down a slash (/).**
3. **Write down the number you divided by.**
4. **You've written down the remainder as a fraction.**

For the example above, that works out to be ⅙ – one-sixth. Everyone gets an extra sixth of a doughnut, just in case they don't have enough sugar already.

I look at fractions in much more detail in Chapter 6.

Working with bigger numbers

Long division seems to be the most feared and dreaded part of numeracy. I understand why: long division is fiddly, tedious and unforgiving – one tiny slip and your answer goes terribly, horribly wrong. But after you get into a routine and you do your sums the same way every time, long division suddenly clicks and you realise you can do it and it's not all bad. So, take a deep breath and say after me: 'long division is just like the division I did earlier in this chapter, only with bigger numbers.'

Just like when you divide by a small number, I suggest you make a times table for the number you want to divide by. If you're not sure how to make a times table, read the section 'Generating a times table' earlier in this chapter – you'll save a great deal of time in the long run.

The layout and process for long division are exactly the same as for short division, which I showed you in the section 'Dividing and conquering' earlier in this chapter.

The only thing that's a little different is that sometimes the left-over number you carry on to the next digit is more than ten. But don't panic – if you wind up with 12 left over and the next number is 9, you just have $^{12}9$, or 129, to deal with next. I show you an example in Figure 4-10.

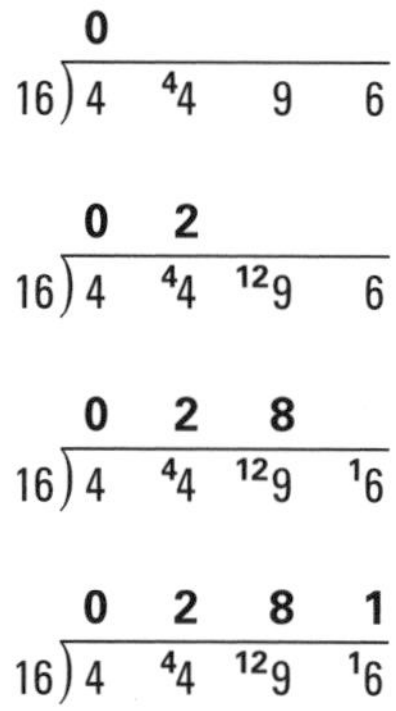

Figure 4-10: Long division – just the same as short division.

Figuring Out Formulas

Fact: the words 'algebra' and 'gibberish' come from the same Arabic root. I don't include much algebra in this book, but you do need to be able to understand how formulas work.

Think of a formula as a very concisely written recipe. For example, if tickets for a concert cost £20 each, plus a £5 booking fee, we can write down the recipe in words as follows:

> *To work out how much the tickets cost, take the number of people going to the concert, times it by 20, and then add 5.*

That's a lot of words, but it works: if you want to buy five tickets, you work out $5 \times 20 = 100$, and then you add 5; $100 + 5 = 105$; so the amount you pay is £105.

In a formula, we give single-letter names to the things we don't know: say, P for the price we have to pay and N for the number of people going. So instead we can write:

$$P = 20 \times N + 5$$

To work out the cost of five tickets as before, you do just the same thing: you replace N (which means 'number of people') with 5. The price is $20 \times 5 + 5 = 105$, just as before.

Wait, those aren't numbers! Looking out for letters

You may be a bit unsettled when you see a formula, especially if a formula contains a lot of letters. The important thing is to keep calm, read the question, and replace the letters one at a time with the numbers they represent.

If and when you go further in your maths studies, you may learn to manipulate letters as if they were numbers, generating and solving all manner of equations. That's beyond the scope of this book – but you can check out *Algebra I For Dummies* if it sounds like your cup of tea.

Missing out the multiply

One of the confusing things about algebra is that normally you don't write a '×' sign to show multiplication (unless it's between two numbers). So instead of writing $P = 20 \times N + 5$, you leave out the '×' and write $P = 20N + 5$. When you talk about objects, you don't say 'three times an egg' – you simply say 'three eggs'. Algebra works exactly the same way.

You may see a little '2' above and to the right of a letter or number, like this: x^2. This is pronounced 'x squared' and just means 'times x by itself.' So, if you know x is seven, you work out $7 \times 7 = 49$. If x is 4, you do $4 \times 4 = 16$. And so on.

Doing the sums in the right order

When you have a complicated formula, doing the sums in the right order is important. Look at the concert example from earlier again:

$$P = 20 \times N + 5$$

If you work out $20 \times N$ and then add 5 you get a different answer than if you do $N + 5$ and then times by 20. If you want to buy tickets for ten people, the first (correct) way would be $20 \times 10 = 200$, and then add 5 to make 205. The other (wrong) way – adding before you times – gives you $20 \times 15 = 300$. Without a specific order for doing sums in, the whole of maths (literally) descends into chaos. Fortunately, a specific order exists. So, when you have a messy formula, here are the steps you take:

1. **Look for any part of the equation set in brackets.**

   Do those sums first. Note that within each bracket, all of the following rules also then apply.

2. **After the bracketed bits, look for any squares – the '2' above the number.**

   Work those out next.

3. **Now look for any times or divide sums.**

   Do these in order from left to right.

4. **Now look for any add or take-away sums.**

   Do these last – again, work from left to right.

Some people remember the order to do sums in a formula using the mnemonic 'BIDMAS' – Brackets, Indices (which means, as far as you're concerned, 'squares'), Divide and Multiply, Add and Subtract.

As an example, if you meet something horrible like:

$$((6 + 13) \times 7 + 5 \times 3) \div 4 \times 100$$

the first thing to do is breathe deeply. Now work out the sums one step at a time. First, look for brackets – there are two pairs. Work out the inner one first: $6 + 13 = 19$. Easy – you can replace that bracket with the number now, to give the following:

$$(19 \times 7 + 5 \times 3) \div 4 \times 100$$

Next up, do the remaining bracket, which contains $19 \times 7 + 5 \times 3$. There aren't any brackets or squares in there, so you do the times-and-divide step. The first times you come to is 19×7 – which you can work out to be 133. The next one is $5 \times 3 = 15$. Replace the sums with their answers to get the following:

$$133 + 15$$

So, that bracket works out to be 148. When you replace the second bracket from before, you have $148 \div 4 \times 100$.

There aren't any brackets or squares in this sum. You have just times and divide to do, and you do these sums from left to right: $148 \div 4 = 37$. Then $37 \times 100 = 3,700$, which is your final answer. Phew!

When you see brackets, they mean 'do this bit first'.

Working out a formula

To figure out a formula, you need two things: the numbers that the letters represent, and the order in which you need to do the sums.

In an exam, you probably won't see too many big numbers in formulas – the examiners are not trying to test your arithmetic with this kind of question. Real life isn't quite as co-operative . . . but in real life you have access to spreadsheets and can let your computer work out the hard stuff.

The process for figuring out a complicated formula is this:

1. **When you see two letters next to each other without a symbol between them, insert a times sign. Do the same thing if you see a number next to a letter.**

2. **Replace each of the letters with the number it represents.**

3. **Work out the sums in the order outlined in the section 'Doing the sums in the right order' earlier in this chapter.**

For example, you might have the formula

$$A = 2fd + fd^2 \div 2$$

and know that $f = 3$ and $d = 4$.

You start by putting in the missing times signs:

$$A = 2 \times f \times d + f \times d^2 \div 2$$

Then you replace the letter f with a 3:

$$A = 2 \times 3 \times d + 3 \times d^2 \div 2$$

Every d becomes a 4:

$$A = 2 \times 3 \times 4 + 3 \times 4^2 \div 2$$

Now your formula contains only numbers. Next, you work through the to-do list. Are there any brackets? Nope. Are there any squares? Yes! You need to work out four squared, which is $4 \times 4 = 16$. You can replace that now:

$$A = 2 \times 3 \times 4 + 3 \times 16 \div 2$$

There are no more squares, so you move on to the times and divides, which you do from left to right: $2 \times 3 = 6$, so $A = 6 \times 4 + 3 \times 16 \div 2$

$6 \times 4 = 24$, so $A = 24 + 3 \times 16 \div 2$

$3 \times 16 = 48$, so $A = 24 + 48 \div 2$

$48 \div 2 = 24$, so $A = 24 + 24$

Finally, you have just an add:

$A = 24 + 24 = 48$

Formulas are involved and can take a while to do. But like many things in maths, if you break down a formula into tiny steps and do those steps in order, you see that formulas are not as difficult as they look.

Chapter 5

Are We Nearly There Yet? Rounding and Estimating

*R*ounding and estimating seem – on first glance – to be much less fundamental to maths than the big four operations you meet in Chapters 3 and 4. I would argue, however – and frequently do – that being able to estimate intelligently is at least as important a maths skill as knowing how to multiply and divide big numbers accurately.

In real life, you have computers for getting accurate answers. I genuinely can't think of a real-life situation where long division would be a useful skill (outside of a maths test). The important thing to me is to be able to ask 'Does that answer make sense, or have I made a mistake in what I've asked the computer to do?'

To answer that question, you need to have an idea of what the answer *ought* to be – which is where estimation comes in. Estimation is a simple process: Round off the numbers in your sum to something easy to work with, and then do the sum with those rough numbers. If I do my estimate and see that my computer's answer is off by a relatively huge amount, I know I've done something wrong and need to check things more carefully. If my answer is in the right neighbourhood, I can feel a bit more comfortable with it. (It doesn't mean I've *definitely* got it right, but it's a good indicator.)

In this chapter, I show you how to round off numbers, and then how to use rounded numbers to produce a solid estimate for your answer. In my opinion, this chapter is probably the most important one in the book for real-life maths (except possibly Chapter 19 on probability).

What's Nearest?

When I was a penniless student, trips to the supermarket were fraught. I had to buy enough food so I didn't starve, but I also had to make sure I didn't spend so much that I got kicked out of my flat for not paying my rent.

Because I thought I was terribly cool (for a maths student, anyway), I refused to walk around with a notepad, diligently adding up the exact cost of every tin of basic value baked beans and loaf of generic textureless bread product I threw in the basket. Instead, I started rounding. 'Those three tins together are 93p – that's about a pound. That cardboard pizza is £1.17 – that's also about a pound. That bottle of student special apple juice is £2.69 – let's call it three pounds.' Instead of juggling long and nasty numbers, I just counted the nearest pounds. For that hypothetical shopping bag, I'd have guessed £5 – not far away from the real value of £4.79, and certainly close enough to say 'I'm not spending too much.'

With rounding, instead of working with the full, ugly number, you pick a number that's near to the ugly one but much easier to work with.

A walk between two towns

Imagine you're walking between two towns, Mored-on-Severn and Leston Eyot, which are a kilometre apart – the same as 1,000 metres. After you walk 100 metres, someone from Leston calls you and says 'Are you nearly there yet?' You say 'No, I'm still closer to Mored.' You reach the 495 metre mark and your phone goes again. 'No,' you say, 'I'm still closer to Mored – but not by much. I'm practically halfway.' At 505 metres, you've gone more than halfway and can say to your friend in Leston 'I'm nearly there – I just passed halfway.' See Figure 5-1.

This is how rounding works: you decide which of the two options you're closer to by looking at whether you're more or less than halfway between the two options.

If you're less than halfway to the next town – or number – you're closer to where you came from (and so you *round down*). If you're more than halfway, you're closer to where you're going (and so you *round up*).

Rounding on a ruler

You don't need to walk between two towns to round off distance to an *appropriate degree of accuracy*. If you measure something with a ruler, you probably automatically do a little bit of rounding – the chances of a measurement landing absolutely precisely on a millimetre mark are pretty small, and in any case you may not need such a precise measurement.

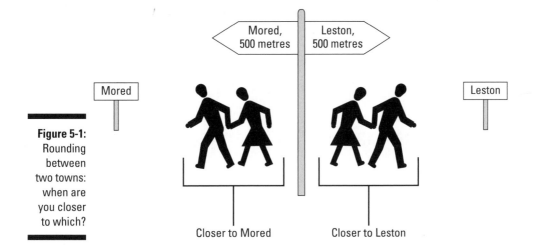

Figure 5-1:
Rounding
between
two towns:
when are
you closer
to which?

Using a ruler, you look and make a judgement about whether the thing you're measuring is closer to one mark or the other. You don't waste time wondering whether the thing is 12.76 millimetres or 12.77 millimetres long – in most cases (where we're concerned, anyway) you can happily say the thing is 13 millimetres long.

You can use a ruler to get the hang of rounding decimals quite easily, at least for numbers up to 30. Here's what you do if you have to round something below 30 to the nearest whole number (I give you a non-ruler-based method later in this chapter, so don't worry if you don't have one to hand):

1. **Find the whole number you want on the ruler.**

   For example, if you want to round 15.43 to the nearest whole number, look on your ruler for 15.

2. **Go to the millimetre mark corresponding to the first digit after the decimal point in your number.**

   In our example, the first number after the decimal point is 4, so you look for the fourth millimetre mark.

3. **Ask whether you're to the left or right of the slightly bigger tick marking midway.**

   In our example, if you're on the 15 side, your answer is 15. If you're closer to 16, you round up to 16. (15.43, for the record, rounds down to 15.)

I show the ruler-rounding process in detail in Figure 5-2.

All rounding is based on the idea of finding which measurement (at a given accuracy) your number is closest to.

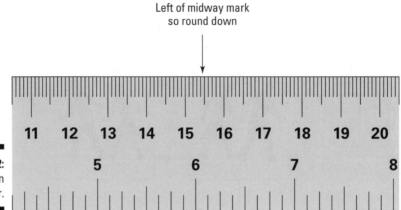

Figure 5-2:
Rounding on
a ruler.

Tie-breaks: What to do when you're midway

When I was about seven, my family went on a day trip to Henley-on-Thames. The river marks the border between Oxfordshire and Berkshire (or, as I liked to think of it back then, civilisation and the barbarian hordes). On the bridge across the river is a marker that says 'Oxon' on one side and 'Berks' on the other. Being a literal-minded elder son and vaguely aware that I was supposed to test boundaries, I stood with one foot on either side of the marker and asked what county I was in. I still believe I deserved a better answer than a tut and a sigh.

In maths though, we have an answer for rounding numbers that are exactly halfway: they go up. If you have to round $4.50 to the nearest pound, you call it five pounds, even though the price is equally far from four pounds and five pounds.

In some fields, the convention is not to round up but to round to the nearest even number. A technical reason exists for this: if you have a lot of tie-breaks, simply rounding up introduces an upward bias to your numbers, and that can be a very bad thing. But unless you work in a lab or other scientific area you probably don't need to worry about this.

A common misconception

The first time I ever realised that a teacher could possibly be wrong, I was about 10. My class was learning about rounding, and the teacher said 'If you have to round something like 1.45 to the nearest whole number, you round it up to 1.5 and then to 2.'

No! No, no, no, no, no! And, furthermore, no.

The number 1.45 is categorically *not* as close to 2 as it is to 1. The number 1.45 is less than halfway to 2, and so you round it down to 1.

The only important number for deciding whether to round a whole number up or down is the one immediately after the dot. Forget everything after that line: those digits are dead to you.

Dealing with Decimal Places

Sometimes, you need to round off numbers to a certain number of decimal places rather than just the nearest whole number. This looks a bit more frightening than whole-number rounding, but the rules are almost exactly the same as for whole numbers. To get a full view of decimal places and how they work, check out Chapter 7.

In the meantime, here's a brief run-down on decimal places:

- When you write a money amount such as £14.99, you put a dot (a *decimal point*) in the number to separate the whole pounds (in this case, 14) from the pence – or parts of a pound (in this case, 99).

- The same thing works for any value that isn't a *whole number* but has something left over. The dot separates the whole numbers (to the left of the dot) from the *decimal part*, which represents parts of a whole number (to the right of the dot).

- As you read from left to right, the number represented by each digit gets smaller – this carries on into the decimal part, so the number immediately to the right of the dot is ten times smaller than the number to the left of the dot, and ten times bigger than its neighbour to the right.

- If we write a number to (say) *two decimal places*, this means the number has two digits to the right of the dot. For example, we usually write prices to two decimal places. *Seven decimal places* means seven numbers to the right of the dot.

- The dot has no magic powers – we simply use the dot as a marker to show where the whole number ends.

Rounding to a certain number of decimal places is less difficult than you may think – but don't worry if you struggle with the concept for a while. Keep practising and you'll get it in the end.

To round a decimal number, you need to think about two things: how many decimal places you have to work with, and whether you have to change the last digit. After you work out those two things, rounding a decimal number is very easy.

Your exam paper will usually tell you how many decimal places to round a number to – for our purposes, the number of places is almost always one, two or three. (To round to more decimal places, the same tricks I show you in this section work just as well.) Here are the steps you need to take – follow along in Figure 5-3 if you'd like to:

1. **Find the dot in the number.**

2. **Count to the right the number of decimal places requested.**

   For example, if you want an answer to two decimal places, count two spaces to the right of the dot. Count between the numbers, so say 'one' after the first digit, 'two' after the second, and so on.

3. **Draw a vertical line after the number you count to.**

4. **Write out the number again, ignoring everything after the line.**

5. **Work out whether you have to round down.**

   Look at the first digit to the right of the line in Step 3. If the number is low (four or below), your answer is the number you wrote down in Step 4.

6. **If the number is high (five or above), you have to round up.**

   This means 'add one to the last digit of your answer'. Write a '1' below the last digit of your answer from Step 4, and add up in columns as normal. What comes out is your answer.

When you have to round up something that ends in a nine, you end up with a zero at the end of the decimal. You might think that's a bad thing or even that you can just get rid of the zero. No: when you round a number, make sure you have as many decimal places as are requested, even if one or more of the digits is zero.

Figure 5-3 shows how to round 16.982 to one decimal place and to two decimal places.

Figure 5-3:   1 6 . 9 | 8 2   to 1 decimal place   1 6 . 9 8 | 2   to 2 decimal places
Two rounding   8 is a high number, round up   2 is a low number, round down
sums.   1 7 . 0 | 1 6 . 9 8 |

Rounding to the nearest penny

Working out a sum to the nearest penny may seem a bit confusing, as there isn't any coin smaller than a penny – but that's exactly why you have to round off money answers.

As with most price-related sums, you have two ways of looking at this: you can think about the number of pence and round to the nearest whole number (which tends to be the easier option, as long as you remember to turn the answer back into pounds at the end); or you can think about the number of pounds and round it to two decimal places. Either way works well – just take your pick.

Rounding to the nearest penny makes perfect sense. Say you buy vegetables at 57p per kilogram and they weigh 450 grams – your veggies total 25.65 pence. But you can't pay 0.65 of a penny. Instead, you round the amount to the nearest penny. Here, that means drawing your imaginary line in the same place as the decimal point, noticing that '6' is a high number and rounding up the last pennies digit up – so the answer is 26p.

Don't confuse 25.65p with £25.65. They'd be very expensive vegetables!

The nearest 10p, and the nearest tenth

Rounding to the nearest 10p and, not coincidentally, rounding to the nearest tenth follow a similar pattern – but with a bit of a 'gotcha' attached. To round to the nearest 10p, draw your rounding line to the right of the number just after the decimal point – that is, the number of 10p pieces you'd need to pay. This is the same as rounding to one decimal place.

Follow exactly the same process as you did before: you look at the number after the line and decide whether it's high or low. If the number's low (zero to four), chop off everything after the line and that's your answer. If the number's high (five to nine), chop off the number after the line and then add one to the last digit. But here's the 'gotcha': if you deal with money, you conventionally write two digits after the dot, even if there aren't any pennies. (I go on a bit of a rant about place values and missing zeros in Chapter 8.) So, don't forget to put a zero on the end if you round to the nearest 10p.

If you have a normal number rather than an amount of money, you shouldn't put a zero on the end, because that makes the number look more accurate than it really is. But don't worry too much about this – nobody's going to yell at you if you mistakenly stick a zero on the end.

The nearest pound, the nearest ten pounds and so on

Rounding to the nearest pound or the nearest whole number is the easiest of all the rounding you can do. You draw your line on the decimal point. Then, as always, round up if the digit after the line is five or bigger, and leave it alone if the digit before the line is four or smaller. Things are similar if you want to round to the nearest ten or the nearest hundred, or whichever bigger number you happen to prefer. The only difference is you need to fill out the number with zeros after you discard 'the rest of the number'. I show you what I mean in the following simple steps – backed up by Figure 5-4:

1. **Draw a line after the digit representing the value you want to round to.**

   If you want to round 324 to the nearest 10, draw the line after the 2 (which is how many tens you have).

2. **Round up or down as normal.**

   If the next digit is four or lower, ignore everything after the line. If the next digit is five or higher, add one to the last digit before the line. In our 324 example, you now have 32.

3. **Add as many zeros as you've thrown away digits.**

   In this case, you add one zero. The final answer is 320.

Figure 5-4 shows you how to round 796.7 to the nearest whole number and to the nearest ten.

Figure 5-4:
Rounding to the nearest whole number and the nearest ten.

796.|7   to the nearest whole number
          7 is a high number, round up
797|

79|6.7   to the nearest whole number
          6 is a high number, round up
80|

          Fill out number with a zero
80|0

When you round to a number bigger than one, make sure you add the right number of zeros on the end – you want your number to be roughly the same size as the number you started with. If you don't add the zero or zeros at the end, you end up with a rotten estimate. The idea of rounding is to come up with a nice round number as close to the number you started with as possible. Thirty-two is nowhere near 324, but 320 is reassuringly close (3,200 would be wrong as well).

That's About Right: Estimating Answers

Imagine you're in a shop and need to buy 80 Christmas cards at 99p each. (I know this example is unrealistic. Just roll with it for a minute.) How much will your cards cost?

Rather than get your calculator or your pen and paper, you can make a sensible guess – each card costs almost exactly a pound, so your total haul is something pretty close to £80. (The exact answer, if you want to practise your arithmetic skills, is £79.20 – so not far off £80.)

Estimating answers like this is one of the most important skills in maths – having an idea of what you expect an answer to be can help you spot calculator mistakes very quickly.

One of the reasons I find estimating so important is that you have to think about what's going on. As an entrepreneur, I'd prefer to employ someone who thinks about problems but sometimes makes mistakes over someone who uses algorithms immaculately. (Everyone makes mistakes. Not everyone takes a moment to check whether the answer is reasonable.) In a multiple-choice exam, estimating is especially useful, as you can often throw out an option or two as completely implausible after a quick estimation.

In the next section, I show you how to come up with a decent estimate of a sum by rounding all of the numbers to the first digit (or *most significant figure*) when asked to, and give you an idea of how to use the technique even when you're not asked about it.

Rough and ready: Rounding to the first digit

The benefit of rounding to the first digit is that you end up with a much easier sum to work with than if you did it the long way. Think about this: would you rather work out $1,234 \times 5,678 \div 60,210$, or $1,000 \times 6,000 \div 60,000$? I know which I find easier – the second one comes out to 100, just using my head. The exact answer is a little over 116, using a calculator – on paper, that question would take *me* several minutes, and I'm good at this stuff.

The trick is to round everything to the biggest number you can get away with. In our example, 1,234 to the nearest thousand is 1,000; 5,678 to the nearest thousand is 6,000; and 60,210 to the nearest ten thousand is 60,000. Then you simply split off the zeros, just like in this example:

A sales rep drives 936 miles in a month. Her car travels around ten miles on one litre of petrol. Petrol costs £1.33 per litre. She estimates her monthly fuel cost by rounding her distance travelled to the nearest 100 miles and the cost of petrol to the nearest 10p.

What is her estimate?

1. **Do the rounding that the question suggests.**

   Round her distance to 900 miles and the cost of petrol to £1.30 (or 130p).

2. **Work out the amount of fuel needed.**

   900 miles divided by 10 miles per litre gives 90 litres of petrol.

3. **To work out the cost, do 90 × 130p.**

   For the moment, put the zeros at the end to one side, but take a note that you have two of them. That leaves you with 9 × 13.

4. **Do this sum using your favourite times method.**

   The answer is 117. Adding the two zeros back on gives 11,700p.

5. **Turn the figure back into pounds.**

   Your final answer is £117.

Take the following steps to round a sum to the first digit:

1. **Take each number in the sum in turn and decide what's the biggest number you can get away with rounding to.**

   Hint: try the first number in the sum that's not a zero.

2. **Round each number, being careful to round up or down as appropriate.**

3. **Work out the sum with the rounded numbers.**

   If the sum gets uncomfortably complicated, check out the section 'Rounding in the middle of the sum' later in this chapter.

4. **Write down your answer.**

Remember your answer is only an estimate. The real answer could be significantly bigger or smaller, especially if you round a number that is close to halfway, or if you round all the numbers in the same direction (all up or all down). The idea isn't to get an exact answer (for that, you can just use a calculator) but to get an answer that's in the right ballpark.

As a rule of thumb, I expect the answer to be between double and half of the estimate most of the time, and practically never more than ten times bigger or less than a tenth of the size – although this varies depending on the calculation and the roughness of the rounding.

Working with all of the zeros

Adding and taking away things with lots of zeros hanging around is easy enough: just make sure the ends of the numbers are lined up before you do the sum as normal.

Multiplying and dividing numbers with zeros at the end is a bit different. Think about the sum $10 \times 10 = 100$. Each of the numbers you multiply has one zero on the end, making two zeros altogether – and it's no coincidence that the answer has two zeros on the end. When you have zeros at the end of a number that you need to times, you end up with all of the zeros at the end of the new number:

1. **To times two numbers ending in lots of zeros together, write down how many zeros are on the end of the first number.**

   If you want to do $30,000 \times 300$, the first number (30,000) has four zeros.

2. **Write down how many zeros are on the end of the second number, and add up the total number of zeros.**

   In our example, the second number (300) has two zeros. Four zeros and two zeros gives six zeros altogether.

3. **Ignore the zeros for a moment and do the times sum with the digits left at the start of the numbers.**

   For example, do $3 \times 3 = 9$.

4. **Add to the end of this number as many zeros as you worked out in Step 2.**

   In our example, add six zeros after 9, to give 9,000,000.

Ooooh! Dividing numbers with loads of zeros

You probably won't have to divide numbers with lots of zeros at the end in your numeracy exam, but here's how to do it:

1. Write the divide sum as a fraction. For example, write $400 \times 200 \div 8,000$ as $(400 \times 200)/8,000$.

2. If the top of the fraction is an add or take-away sum, work out the sum. This is surprisingly important. You can leave a times sum alone though.

3. If you have a number on top that ends in a zero and a number on the bottom that ends in a zero, cross out one of each.

4. Keep doing this until you run out of zeros on top or bottom.

5. Now do the sum. It should be easier now.

What you're really doing here is cancelling a factor of 10 from the top and bottom of a fraction. I go into cancelling in more detail in Chapter 6.

Rounding in the middle of the sum

Sometimes, halfway through a sum, you wind up with a horrible mess of a number. Maybe you add things together and end up with something that isn't so nice any more. Maybe you times or divide and get an ugly answer.

Don't panic! My swift and brutal solution to your problem is simply to round off again.

Make sure you follow the same steps as before – remember, you're looking for a rough ballpark answer rather than getting a bang-on result.

If you're confident, you can save time by working out only the first couple of digits of a sum and how large it ought to be. For example, if you want to do $10,000 \div 7$, you can work out the first two digits (1 and 4), realise they're in the thousands and hundreds places (respectively) and say 'My answer's about 1,400 – I'll round that to 1,000.' Of course, if you prefer to work out the whole sum, go ahead. But don't fall into the trap of thinking your answer's accurate just because it's got lots of decimals – you've already done some rounding, so the answer's still an estimate.

Careful with that calculator!

Calculators are marvellous machines. Even since I was at school, which wasn't *that* long ago, calculators have become more and more sophisticated – and I'm pretty sure that the scientific calculators on the market today are more powerful than the best computers I had access to in the early 1990s.

As far as *Basic Maths For Dummies* goes, you don't need a crazy fancy calculator – one that doesn't have much more than the numbers, the operations (add, take away, times, divide) and a decimal point should serve you perfectly well. But if you hope to do more maths after reading this book, you may want to invest in a good scientific calculator. My personal favourite is the Casio FX-85 (the one with a big round button near the top). It's even available in pink, if you think black and grey are too boring!

Whatever kind of calculator you have, it's certain to be better at doing sums than you are. I don't mean that as a put-down – after all, calculators are better at pretty much *any* kind of sum than *I* am – but a teensy-weensy problem exists with all calculators.

The problem with learning to juggle is that the balls always go where you throw them. Likewise, calculators do *exactly* what you tell them. Your calculator has no idea whether you've asked it to do the right thing.

This is one of the reasons I recommend coming up with a rough answer on paper, even if you use a calculator: your estimate can save you from coming up with embarrassingly wrong answers.

In real life, do a rough version of every sum on paper before you ask your calculator or computer, just to get a ballpark figure for the answer.

Checking your answers

One of the main uses of estimating and rounding is to check that the answers you get from more accurate methods – say, working out sums on paper or using a calculator – make some kind of sense.

Estimating and rounding can't tell you for sure that you have the right answer – if it could do that, there wouldn't be any point doing the full sum – but the various methods I show you in this chapter give you an idea of whether your accurate answer is reasonable.

The first, and probably most reliable, port of call for estimating an answer is to think about your real-life experience and guess what range of answers you expect to get. For example, to answer a question about a car journey between two towns 50 miles apart, try thinking about the last time you made a journey of about 50 miles. How long did your journey take? Depending on the roads, maybe an hour or two? Here's how I may react to some different answers:

- ✔ **My guess and my answer disagree significantly:** If I work out an answer of 12 minutes or 8 hours, I may scratch my head and say 'That can't *possibly* be right!' and do the sum all over again.

- ✔ **My guess is a little off of my answer:** If I get an answer of 40 minutes or 2½ hours, I may say 'Hmm, that's not quite what I expected, but it's not completely implausible – let me check my numbers just to be sure.'

- ✔ **My guess is pretty good:** If I get an answer somewhere between an hour and two hours, I may pat myself on the back and say 'That seems about right – I can run with that.'

This is pretty much the standard way of checking your answers, whatever method you pick to come up with a ballpark number. You can decide whether your answer is just plain wrong, a little iffy, or perfectly plausible – and do the appropriate thing (start over, check it over or celebrate).

Part II
Parts of the Whole

In this part . . .

Everything in this part is adding, taking away, timesing and dividing. It's very important that you remember that, because those things are easy.

The only hard thing about using them to work with fractions, decimals, percentages and ratios is knowing which one to use and when. In this part, I give you recipes for how to figure out this kind of sum, and introduce you to the Table of Joy – a method you can normally use for a third to a half of the questions in an exam.

(See? I slipped in some fractions there and you barely noticed.)

Chapter 6

Cake or Death: Fractions without Fear

*E*ddie Izzard has a routine about how the Spanish Inquisition could never have happened in England. The Spanish Inquisitors offered a choice between repenting and dying. In England, they'd ask 'Cake or death?'

Somehow, people who ask questions about fractions are viewed with only slightly less fear and suspicion than the Spanish Inquisition. Given a choice between a fractions question involving cake, and death, I'm not sure everyone would pick cake.

In this chapter, I try to turn the question of cake or death into a no-brainer. Fractions are the topic in maths where most students begin to struggle. I fully understand your fear – fractions aren't an intuitive topic for most people. So I take things very slowly and carefully: I start with fractions you know about and are comfortable with, from the worlds of time and pubs, before I explain what fractions are and some of the things you can do with them.

I show you that two different fractions can be the same, and I help you work out which of two fractions is bigger.

I also show you how to add and take away fractions and how to use a calculator to help you with fractions.

After that, you can go and grab a slice of cake.

Familiar Fractions

You may protest that you've never 'got' fractions – which is entirely normal, because fractions are confusing and catching up once you're confused is difficult. But do me a little favour: for however long you take to read this chapter, try to forget that you don't like fractions. Give fractions another chance. Start from scratch and get your relationship with fractions off on the right foot this time. Fractions are not as horrible as you think. You probably even use some fractions without even noticing.

For example, you may watch a football match and say 'It was only a half-chance, but he'd normally score that nine times out of ten,' just as the ref blows for half-time and you tuck into a quarter-pounder. That sentence is full of fractions that aren't at all frightening (except for the burger).

No half-measures

Readers with a low tolerance for bad jokes should skip the next paragraph. Frankly, I'm surprised you've made it this far.

So, a man walks into a bar and orders a half-pint. The bartender says, 'Is this some kind of joke?'

Fine, I'll get my coat. Where was I going with that? Oh, yes, the half-pint. Two half-pints are the same as a whole pint – which you probably know. In fact, two halves of anything is the same as a whole one. Two half-pizzas are the same as a whole pizza. Two halves of a football match make up a whole game.

The only exception that springs to mind is that two half-baked ideas don't make a wholly baked idea.

Time to split

You use fractions when you talk about time all the time. I bet you say 'half past four' rather than 'thirty past four'. Quarter-hours (15 minutes) and half-hours (30 minutes) are so deeply ingrained in your language that you probably don't even notice that you use fractions when you tell the time. Figure 6-1 shows how these fractions fit together.

Figure 6-1:
Fractions
on a clock
face.

Two half-hours make up a whole hour. Four quarter-hours also make up a whole hour. Two quarter-hours make up a half-hour. And – this may seem obvious – three quarter-hours make up three-quarters of an hour. These sums work the same for anything else as they do for hours. Here are the sums written out:

- $2 \times \frac{1}{2} = \frac{1}{2} + \frac{1}{2} = 1$
- $4 \times \frac{1}{4} = 1$
- $2 \times \frac{1}{4} = \frac{1}{2}$
- $3 \times \frac{1}{4} = \frac{3}{4}$

Sizing Up Fractions

A fraction is two numbers, one written above the other – like this: $\frac{1}{2}$ – or with a slash between them – like this: 1/2. We use fractions to describe part of a whole number – a half is less than a whole.

Throughout this chapter, I talk about the top and the bottom of the fraction. If you read other maths resources as well, you may also see the terms *numerator* (top number) and *denominator* (bottom number).

One of the hardest things about fractions is deciding whether a fraction is large or small. If you ask me whether $5/7$ or $2/3$ is bigger, I'd take at least a few seconds to figure out the answer – and I'm supposed to be good at this. (In fact, $5/7$ is bigger, but not by very much. I show you how to figure out this sum later in the chapter.)

Sometimes fractions seem completely counterintuitive: you may ask 'How on earth can $1/2$ be bigger than $1/3$, when three is bigger than two?' In this section I try to answer that question – and then I try to convince you that fractions are actually quite logical if you think about them in a particular way.

The bottom of the fraction: How big is your slice?

A half, which has a two on the bottom of the fraction, is bigger than a third, which has a three on the bottom. Imagine a cake. (Death is not an option in this section, sorry.) Imagine cutting the cake into two equal pieces: each piece is then a half – $1/2$. If you cut the cake into eight pieces, each piece is an eighth – $1/8$. And if you cut the cake into a million pieces, each piece is a millionth.

The number on the bottom of the fraction is how many slices of that size you need to make up a whole cake, or how many numbers of that size you need to make up a whole one.

The bigger the number on the bottom, the more slices you need and the smaller the slices have to be. If you have two fractions with the same top number, the bigger fraction is the one with the smaller number on the bottom. For example, $1/4$ is bigger than $1/5$.

The top: How many slices?

The top of the fraction works exactly the way you may expect. Three-quarters of a cake just means 'Split the cake into four equal slices, and take three of those slices.' (Three-quarters of a cake is three times as much cake as a quarter of a cake.)

The bigger the number on the top, the more slices you have. A bigger number on top means more slices and thus more cake. If two fractions have the same bottom number, the one with the bigger number on top is the bigger fraction.

For example, $\frac{3}{4}$ is bigger than $\frac{1}{4}$. Look at Figure 6-2 if you don't believe me.

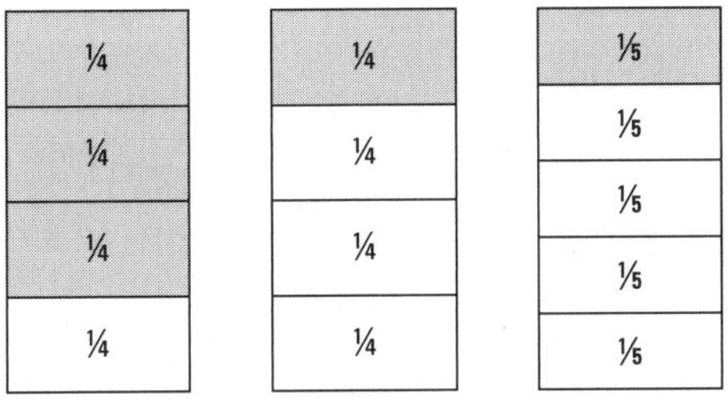
1/4
1/4
1/4
1/4
1/4
1/4
1/4
1/4
1/5
1/5
1/5
1/5
1/5

Figure 6-2: Comparing sizes of slice: $\frac{3}{4}$ is bigger than $\frac{1}{4}$, which is bigger than $\frac{1}{5}$.

The number in front

Sometimes you see something like $1\frac{3}{4}$. This means 'one and three-quarters'. Thinking of pudding again, this means a whole cake plus three-quarters of another cake. The number in front is how many full cakes you have.

If you have $6\frac{1}{4}$ cakes and I have $2\frac{7}{10}$ cakes, it doesn't matter that my fraction is bigger than yours – you have much more cake than I do, because your number in front (6) is bigger than my number in front (2). This is similar to money: if you have £6.25 and I have £2.70 – the pennies (parts of a pound) are less important than the full pounds, and you are considerably richer than I am. (Get the low-down on all things money in Chapter 11.)

Cancel That!

Baseball legend Yogi Berra was as famous for his turns of phrase as he was for his skills. One story involves a waitress asking whether he'd like his pizza cut into four or six slices. Berra replied, 'Oh, four please. I don't think I could eat six.'

The joke here is that the amount of pizza he ate wouldn't change – he'd eat fewer slices, but they'd be bigger. He'd eat four quarters of a pizza or six sixths ($\frac{4}{4}$ or $\frac{6}{6}$) – either way, it's a whole pizza, as I show in Figure 16-3.

So, two fractions may be the same even if they look different.

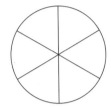

Fractions in disguise

Two fractions that look very different can actually be the same thing. For example, ¼ is the same as ⁶⁄₆, which is the same as 1. In fact, any fraction where the top is the same as the bottom will give you 1 – so ⁵⁄₅ is the same as ²⁵⁄₂₅ and ¹⁸⁷⁄₁₈₇ – and they're all equal to 1.

If you look at Figure 6-3, you see that half of Yogi's pizza (½) is the same thing as two of the quarter-slices (²⁄₄) or three of the sixth-slices (³⁄₆). In all of these cases, the bottom number is double the top number. This rule always applies – so you can also say that ¹⁰⁄₂₀ and ⁵⁰⁄₁₀₀ both equal a half.

Making cancellation easy

Two fractions are equal to each other if they *cancel down* to the same fraction. Cancelling down – also called writing a fraction in its simplest form – is all about making the numbers in a fraction as small as possible. It's much clearer what you mean if you say 'a half' instead of ⁴⁸⁄₉₆.

Cancelling down a fraction to its simplest form is easier than you think. I'd even go so far as to say cancelling is a snap.

A whole lot of nothing

If you have a fraction where the top and bottom both end in zero, the first cancellation step is as easy as things get: you cross off the last zero from both numbers. Keep doing this until you hit a number that doesn't end in zero. So, the fraction ⁴⁰⁰⁄₂₈,₀₀₀ is the same as ⁴⁰⁄₂,₈₀₀ and also equals ⁴⁄₂₈₀. We can't get rid of the 0 at the end of 280 because the top number no longer ends in zero.

This idea works because if you divide (or times) the top and bottom of a fraction by the same number, you end up with the same fraction. To remove a zero, you simply divide top and bottom by ten and make the fraction simpler.

Playing snap with factor pairs

The good news is you can divide by any number you like, so long as both of the numbers let you. The idea is to find factor pairs of the number on top and the number on the bottom. A *factor pair* is two numbers that times together to make another number. For example, 12 has several factor pairs: 1 and 12; 2 and 6; and 3 and 4. For cancelling fractions, you only need pairs that don't include 1. If your number has no factor pairs (in which case, the number is a *prime number*), don't worry – you just don't split that number up.

Part of the beauty of using factor pairs is that the pair you pick doesn't matter – if you go on for long enough, you find the simplest fraction. If you want to sound super-brainy when explaining this, be sure to throw around the phrase 'the fundamental theorem of arithmetic'. People will probably stop asking you questions after that.

Here's the method for using factor pairs to cancel down a fraction:

1. **If you can, split any top number into a factor pair.**

2. **If a number on the top matches a number on the bottom, say 'snap' and cross them both out.**

3. **If you can, split any bottom number into a factor pair.**

4. **If a number on the top matches a number on the bottom, say 'snap' and cross them both out.**

   If you run out of numbers on the top or the bottom, write a 1 to fill the gap.

5. **Repeat Steps 1–4 until you can't find any more factor pairs.**

6. **Times together any numbers on the top.**

   This is your cancelled down top.

7. **Times together any numbers on the bottom.**

   This is your cancelled down bottom.

In Figure 6-4 I show an example of cancelling down $\frac{4}{20}$ to give $\frac{1}{5}$.

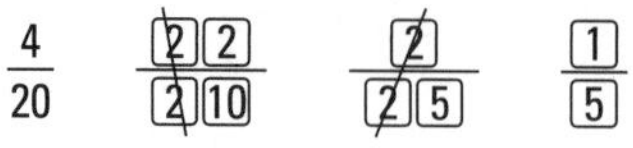

Figure 6-4:
Cancelling
down ¹⁄₂₀.

$$\frac{4}{20} \qquad \frac{\cancel{2}\,\cancel{2}}{\cancel{2}\,10} \qquad \frac{\cancel{2}}{\cancel{2}\,5} \qquad \frac{1}{5}$$

Doing Sums with Fractions

You can think of a half as one cake divided into two pieces. A half – ½ – is the same thing as $1 \div 2$. Three-quarters – or ¾ – is the same thing as $3 \div 4$. This pattern is true for any fraction: ⁵⁶⁄₁₀₃ is the same as $56 \div 103$ (and no, I don't need you to work out that sum!). This explains why a half is the same as three-sixths or fifty-hundredths: if you work out the sums $1 \div 2$, $3 \div 6$ and $50 \div 100$, you get 0.5 or ½ each time.

Writing a number as a fraction of another

Exam questions often ask something like 'What is 12 out of 30 as a fraction?' This simply means work out $12 \div 30$ – which is just the same as ¹²⁄₃₀, which you can cancel down using the 'snap' method from earlier. You get ⅖.

A variation on this question is 'Which fraction is closest to . . .?' For example, approximately what fraction is 5,843 out of 28,946? This is a horrible thing to ask anyone, but here's what to do when the examiner uses words such as 'approximately', 'roughly' and 'nearest to' in a fractions context:

1. **Round the top number as roughly as possible.**

   In this case, the number is 5,843. Rounding to the nearest thousand seems good, so call it 6,000.

2. **Round the bottom number in a similar way.**

   The roughest sensible rounding is to the nearest ten thousand, which would be 30,000 in our example.

3. **Cancel down the rounded top over the rounded bottom.**

   In this case, using the zeros trick or the snap method with 6,000/30,000 gives ⅕.

The same size of slice: Adding and taking away fractions

Imagine you find two-thirds of a pizza in the fridge. You only feel like eating half a pizza, so you cut off the excess, put the rest on a plate and wonder how big the slice you've left for tomorrow's lunch is.

You write down the sum: $\frac{2}{3} - \frac{1}{2} =$ . . . and then you find you have a problem. If you do it the way that seems obvious – do 2–1 to get the top and 3–2 to get the bottom – you end up with $\frac{1}{1}$, and you know there's not a whole pizza left in the fridge. The pizza was initially cut into six slices – and four of them were left. $\frac{4}{6}$ is the same as $\frac{2}{3}$. Then you picked up three of the slices ($\frac{1}{2}$ is the same as $\frac{3}{6}$), to leave only one. So $\frac{2}{3} - \frac{1}{2}$ is the same as $\frac{4}{6} - \frac{3}{6} = \frac{1}{6}$.

To add and take away fractions, the bottom number – the size of the slice – must be the same on both fractions.

My method above works just as well if the pizza had been cut into 12 rather than 6 slices. Two-thirds is the same as $\frac{8}{12}$, and a half is the same as $\frac{6}{12}$, so you have two slices left. Two-twelfths, when you cancel down, is the same as $\frac{1}{6}$ – the same answer as before.

For all you people (like me) who prefer pictures to words, I give a visual representation of the pizza in the fridge in Figure 6-5.

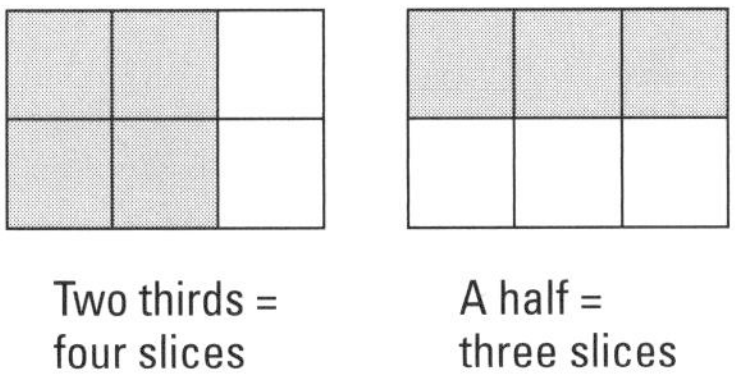

Figure 6-5: Taking away fractions using the same size of slice. Two-thirds of a pizza is the same as $\frac{4}{6}$; half a pizza is the same as $\frac{3}{6}$. Taking away the slices leaves you with $\frac{1}{6}$.

Here's a way to reliably and quickly add or subtract any fractions you happen to meet:

1. **Write down the two fractions you want to add or take away.**

   Leave plenty of space between them.

2. **You need to times the top and bottom of the first fraction by the bottom of the second fraction.**

   Write a × sign and the number from the bottom of the second fraction next to the top and bottom of the first fraction. Have a look at Figure 6-6 to see what I mean.

3. **Now you need to times the top and bottom of the second fraction by the bottom of the first fraction.**

   Write a × sign and the number from the bottom of the first fraction beside the top and bottom of the second. I show this in Figure 6-6.

4. **Work out the times sums and write them on the line below.**

   You should end up with two fractions, with the same bottom number.

5. **To do an add sum, add the tops together; to do a take away sum, take the tops away.**

   The answer to this sum is the top of your final fraction.

6. **For the bottom of the final fraction, use the number you created for the bottom of both fractions.**

7. **Cancel the fraction down if you need to.**

   Turn the number into a mixed fraction if the top is bigger than the bottom.

Figure 6-6:
A take-away and an add sum.

$$\frac{1}{2} - \frac{1}{6} = \frac{1 \times 6}{2 \times 6} - \frac{1 \times 2}{6 \times 2} = \frac{6}{12} - \frac{2}{12} = \frac{4}{12} = \frac{1}{3}$$

$$\frac{2}{5} + \frac{1}{7} = \frac{2 \times 7}{5 \times 7} + \frac{1 \times 5}{7 \times 5} = \frac{14}{35} + \frac{5}{35} = \frac{19}{35}$$

Calculating fractions of a number

You may need to work out a fraction of a number – for example, $\frac{3}{5}$ of 100. This seems terrifying at first, but you actually do a very simple sum: you just times the number by the top of the fraction and divide by the bottom of the fraction – or you can divide by the bottom first and then times by the top:

1. **Divide the number by the bottom of the fraction.**

   In our example, you do $100 \div 5 = 20$.

2. **Times the result of Step 1 by the top of the fraction.**

   In our example, you do $20 \times 3 = 60$. So $\frac{3}{5}$ of 100 is 60.

Comparing fractions

To work out which of two fractions is bigger, you play with the fractions until they have the same number on the bottom. Here's my recipe:

1. **Times the top and bottom numbers of the first fraction by the bottom of the second fraction.**

2. **Times the top and bottom numbers of the second fraction by the bottom of the first fraction.**

   You now have two fractions with the same bottom numbers – which means the individual slices are the same size so you can compare them.

3. **Compare the tops of the two fractions.**

   The bigger fraction is the one with the bigger top. In terms of cake, the fraction with the bigger top contains more slices.

Fathoming Fractions on Your Calculator

One of the most useful tools in the real world for dealing with fractions is your calculator.

You may not be allowed to use a calculator in an exam – always check beforehand – but you can definitely use a calculator when you practise maths at home.

If you have a scientific calculator, it can probably handle fractions for you. Even calculators that don't handle fractions in the obvious sense are useful – you can do the same sums using decimals, which I tell you all about in Chapter 7.

In this section I run you through the ins and outs of using a calculator with fractions. You may want to turn to Chapter 7 as well, where I cover using a calculator with decimals.

Using the fraction button

Whoever decided to put a fraction button on calculators will be given a knighthood when I rule the world. The fraction button has made a whole area of maths much more accessible and useful and I blow a giant raspberry in the face of anyone who grumbles about dumbing down.

On my calculator the fraction button is towards the top left, underneath 'Abs'. The picture on the button looks like a filled white box on top of a line, on top of an empty white box. If your calculator doesn't have a fraction button, don't worry – just skip straight to the section 'Doing fractions with decimals' later in this chapter.

No matter how many times you press 'Abs', the button won't help you get a six-pack. Equally, 'tan' does nothing at all to make you look like you've been somewhere nice on holiday (in fact, 'Abs' and 'tan' are abbreviations for 'absolute value' and 'tangent', functions used in more advanced maths, which you don't need to worry about for now, phew).

You use the fractions button a little differently from the more everyday buttons on your calculator. Here, and in Figure 6-7, I walk you through how to do a sum like $\frac{3}{5} \times \frac{7}{8}$:

1. **To enter a fraction, press the fraction button.**

   The calculator shows a couple of stacked boxes in the display. The top box may contain a flashing line.

2. **Type the top number of the fraction.**

   In our example you press 3.

3. **Press the round button.**

   The flashing line moves into the lower box.

4. **Type the bottom number of the fraction.**

   In our example you press 5.

5. **Press the right side of the round button to escape from the box.**

6. **Carry on with the rest of the sum.**

   In our example you press the × button and enter $\frac{7}{8}$ using the fraction button like you did for Steps 1–5.

7. **Press the = button.**

   The calculator shows you the answer: $\frac{21}{40}$.

If you need to turn your answer into a decimal, press the button marked S↔D. On my calculator this button is in the middle towards the right, just above the red 'DEL' button.

Figure 6-7:
Typing
fractions
into a
calculator.

$$\boxed{} \frac{\boxed{}}{\boxed{}} \quad \boxed{1} \frac{\boxed{}}{\boxed{}} \quad \boxed{1} \frac{\boxed{3}}{\boxed{}} \quad \boxed{1} \frac{\boxed{3}}{\boxed{4}}$$

If you need to type in something like 1¾ into the calculator, it's pretty similar – the only difference is you press 'shift' before the fraction button:

1. **Press 'shift' (in the very top left) then the fraction button.**

   You see three boxes – a big one on the left and two smaller ones stacked as before. The big box is for the number in front, and the smaller ones for the fraction, just like before.

2. **Fill in the first box with the whole number.**

   In our example, this is 1.

3. **Press right on the round button and fill in the top.**

   Three in our example.

4. **Press down on the round button and fill in the bottom.**

   Four in our example.

5. **Press right on the round button to escape from the boxes.**

   Now carry on with the rest of the sum.

Don't worry if you make a mistake. Just press the 'AC' or 'C' button to clear everything away and start again. Or, if you feel brave, use the arrow keys and delete button to try to fix the mess.

Chapter 7 covers using a calculator with decimals.

Doing fractions with decimals

If your calculator doesn't have a fractions button, or you don't like the thought of using this button, you can work out your fraction sums using decimals. Instead of using the fraction button, you need to turn your fractions into divide sums. Here's an example for the sum ½ + ⁹⁄₁₀:

1. **Type the top of the first fraction, then the divide key, then the bottom.**

   For this example, you do 1 ÷ 2.

2. **Carry on with the sum in the normal way.**

   In our example, type the + sign.

3. **Treat the second fraction as you did in Step 1.**

   In our example, type 9 ÷ 10.

4. **Press the = key.**

   Your calculator should say 1.4. If your calculator gives the answer as a fraction – in this example, 1 ⅖ – press the S↔D button in the middle towards the right to convert the fraction to a decimal.

A recurring theme

I start this section with a divide sum: what is 1 ÷ 3? (If you find this difficult, look first at Chapter 4, where I give the details of division.) For example:

I have £1 and want to split it between three people. I split the pound into ten 10p pieces and give everyone three 10p pieces. But I have a 10p piece left over. I change the 10p piece into ten 1p pieces, and I give everyone three 1p pieces. But I have a penny left over. Because I'm in charge, I decide a tenth of a penny exists, and I change my penny into ten tenth-of-a-penny coins. Again, everyone gets three . . . and again, I have one left over. No matter how finely I divide the coins, everyone always gets three coins and I have one left over. I show what I mean in Figure 6-8.

Figure 6-8:
1 ÷ 3: a
recurring
decimal.

$$0 \;.\; 3\;\; 3\;\; 3\;\; 3\;\; 3 \ldots$$
$$3\,\overline{)\,1\;.\,{}^1 0\;{}^1 0\;{}^1 0\;{}^1 0\;{}^1 0}$$

This pattern repeats forever. We call the number a *recurring decimal* because the number three recurs over and over and over and . . .

Every fraction with a whole number on the top and bottom can be written either as a terminating decimal or as a recurring decimal. Sometimes the recurring pattern is longer than one digit – for example, $\frac{1}{11}$ = 0.0909 . . . Sometimes the pattern doesn't start straight away – for example, $\frac{1}{6}$ = 0.16666 . . . Eventually the number either falls into a pattern or stops – in which case the number is a *terminating decimal*.

Recurring decimals are particularly problematic on a calculator that doesn't deal comfortably with fractions. If you work out the sum $\frac{1}{2} + \frac{1}{3} = \frac{5}{6}$ and want to your answer on a calculator, figuring out what to type in is difficult – try following these steps to see what I mean:

1. **Work out the first number on the calculator.**

   In our example, do 1 ÷ 2 and write down the answer (0.5). If the answer's a recurring decimal, just write the first few digits until you see a pattern.

2. **Do the same for the second number.**

   In our example, work out 1 ÷ 3 and write down 0.333 (no need to go overboard with the 3s).

3. **Type the sum into the calculator with decimals.**

   For our example, do $0.5 + 0.333$ and write down 0.833.

4. **Calculate the answer you worked out by hand.**

   $5 \div 6 = 0.8333 \ldots$

5. **If the numbers are pretty much identical, be happy you have the answer right.**

   Don't worry too much about the last decimal place of what you write down, but any differences before that might be a clue that something is wrong.

Chapter 7

What's the Point? Dealing with Decimals

. . . .

In This Chapter

▶ Discovering decimal points

▶ Putting numbers in their place

▶ Having fun with fractions and decimals

▶ Doing sums with decimals

. . . .

A lot of capable, intelligent maths students still shudder a bit when they see one of those 'funny dots' in a number. These people are perfectly capable of using the 'funny dots' when they deal with money, but they freeze up completely when they see the dots in any other context.

In this chapter I show you how the funny dots – or, rather more correctly, decimal points – work in a money context.

I also show you how decimals are linked in with fractions – so you may choose to read this chapter in conjunction with Chapter 6, where I cover fractions in detail. I promise I'm gentle in this chapter – and I promise decimals are much easier than you may think.

Also in this chapter I cover the history of place values to show you why decimals are nothing to be afraid of. You can then use that knowledge to lay out add and take-away sums with dots, without going dotty.

I show you how to multiply and divide with decimals, and quickly run through a few of the tricks and pitfalls that crop up when you use a calculator to do this kind of thing.

A Dot You Know: Decimals and Money

Look at a receipt or a price tag or a bill. Almost any time you see an amount of money written down, it has a dot in it. That dot is a *decimal point* and works just the same in numbers as it does in pounds and pence.

In the context of money, the dot shows where the pounds end and the pence begin. The numbers before the dot are the number of whole pounds you spend, and the numbers after the dot are the number of pence you spend.

Put differently, the dot divides the amount into whole numbers on the left, and things less than a whole number on the right.

As you go left from the dot, the value of each number gets bigger: the first digit to the left of the dot is pounds, the second digit to the left is tens, the third digit to the left is hundreds, and so on. Each digit is ten times bigger than the one to its right.

Put another way, as you move right through the number, the value of each digit is ten times smaller than the one to its left. This continues through the decimal point – so the first digit after the dot is ten times smaller than the one to the left of the dot, and the next digit is ten times smaller still.

Looking before and after the dot

The numbers to the right of the dot confuse many maths students. Humans are generally much better at thinking about whole things than parts of things – and unfortunately for whole-number thinkers, the things to the right of the dot are less than a whole.

In money terms, the first digit after the dot is the number of 10p coins you need, and the second digit is the number of pennies.

You need ten 10p pieces to make up a pound, so a 10p piece is ten times smaller than a pound. You need ten 1p pieces to make 10p – so a 1p piece is ten times smaller than 10p. So each place further you go after the dot is worth ten times less than the one before.

Missing off the last zero

Decimals sometimes end unexpectedly, especially if you do decimal sums on a calculator or computer. To see what I mean, get a scientific calculator and type in '25 ÷ 2'. If the answer is a fraction, hit the S↔D button above the red button. The calculator should give the answer 12.5.

Many people look at this answer and think 'If I split £25 between two people, they each get £12 and 5p.' But if you hand out £12 and 5p twice, you only give away £24 and 10p.

What 12.5 actually means is £12.50 – but your calculator hasn't shown the last zero. Why would it be so lazy?

As far as the calculator is concerned, it only tells you about the coins and notes you really need. When the calculator gives the answer 12.5, it tells you that you need one £10 note, two £1 coins, and five 10p pieces. The calculator doesn't tell you about the 1p pieces you don't need – or likewise about the £100 and £1,000 notes you don't need.

Sometimes the calculator gives a zero in the middle of a number – for example, in the amount £1.07. In this case, the zero is being used as a space – but I explain that in detail in the next section.

A Whole Lot of Nothing: Place Values and Why They Matter

When I was at primary school, I spent weeks playing with blocks of wood – 10-by-10 squares representing hundreds, rows of 10 representing 10, and loose cubes representing ones.

At the time, I simply had fun playing with the blocks, and occasionally a grown-up asked me to do a sum with the blocks. But much later I realised the teachers were trying to relate the sums I did on paper to the blocks.

When you do a sum using blocks that involves a single 100-block, a single 10-block or a single 1-block, you write down the digit 1. You can't tell by looking at the 1 in isolation how big your number is. To find out the size of your number, you look at the numbers around the 1 – particularly how many spaces the 1 is from the end of the number or from the decimal point. If you have, say, three blocks of any size, writing down the digit 3 doesn't tell you anything about the size of the blocks.

For the rest of this section, I assume all numbers have a decimal point. If you have a number that doesn't have a decimal point, just put a dot at the very end.

Hundreds, tens and units: Looking left of the decimal point

The number immediately to the left of the decimal point tells you how many *units* – £1 coins, say – you have. The next digit to the left tells you about things ten times bigger – in this case, £10 notes. The next digit to the left counts things ten times bigger still – so £100 notes. You can work like this forever: each digit tells how many 'banknotes' you have of a particular size, and each space you move to the left tells you about notes that are ten times bigger than the one before.

The flip side is that as you move to the right, things get ten times smaller for each space you move.

Tenths, hundredths and smaller: Roaming right of the decimal point

If you have read some of the previous paragraphs, you may guess that moving one place to the right of the decimal point is ten times smaller than the unit just to the left of the decimal point. If you split a £1 coin up into ten equal pieces, you get ten tenths of a pound, or ten 10p pieces. The first digit after the dot tells you about tenths.

By the same token, the digit after that tells you about hundredths – ten times smaller still. The pattern goes on for ever: the third digit is how many thousandths, the fourth how many ten-thousandths, and so on.

The further to the left a digit is, the more *significant* it is – the larger the value it represents. Every digit is ten times bigger than the one to its right and ten times smaller than the one to its left.

In Figure 7-1 I show you how the numbers fit together.

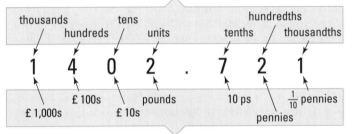

Figure 7-1:
Place
values,
including
decimals.

Zeroing in on zeros

We use Roman numerals all over the place – to give the date at the end of TV programmes, in the names of kings and queens, on clocks . . . I'm sure you can think of half a dozen other places as well. The Romans used an odd system of letters to represent numbers, so that MMXIII is 2,000 (two Ms) plus 10 (one X) plus 3 (three Is), making 2,013.

The Roman system isn't bad for counting. If you don't need any hundreds – as in the example above – you just don't write the hundreds in the number.

In the Arabic system used almost everywhere in the world today, you can't just miss out digits when there aren't any banknotes of that value. If you do that, you can't tell the difference between 20 and 2, or between 100, 10 and 1. You don't know if the London Olympics are in 212, 2012, 2102 or 2120. If you have thousands but no hundreds, you put a zero in the hundreds place – which is between the thousands and the tens. This works in much the same way to the right of the decimal point, except you go the other way: if you have hundredths but no tenths – for example, £12 and 7p – you put a zero in the tenths column so you can tell what value each digit has. You write twelve pounds and seven pence as £12.07 – not £12.7, which is actually £12.70.

Linking Decimals to Fractions

In this chapter I talk about the number directly after the dot as 'how many tenths', the number after that as 'how many hundredths' and so on. Decimals are just another way of writing fractions in a simple way. You may want to have a look at Chapter 6 before you wade through this section if you're not comfortable with fractions. The number 75.43 means seven tens, five units, four tenths and three hundredths – or $70 + 5 + \frac{4}{10} + \frac{3}{100}$. Each new decimal place gives you another fraction with an extra zero on the bottom – the next one here would be something over 1,000, then something over 10,000, and so on.

In practice, we don't work out decimals like that. Instead, you know that $\frac{4}{10}$ is the same as $\frac{40}{100}$, so $\frac{4}{10} + \frac{3}{100}$ is the same as $\frac{40}{100} + \frac{3}{100}$, or $\frac{43}{100}$. I describe this nice easy way to turn decimals into fractions in detail in this section.

Turning a fraction into a decimal is a bit more of a hassle. You first turn the bar of the fraction into a divide sign. So to work out $\frac{3}{5}$ as a decimal, you work out $3 \div 5$ – which is 0.6. I show you how to do this later in this section.

Also in this section I include a table of decimal/fraction conversions with some notes on remembering them easily.

Converting decimals to fractions

Converting decimals to fractions is as easy as counting. Well . . . maybe not quite as easy as counting, but counting is a significant part of the sum you do. Here's what you do:

1. **Ignore anything to the left of the decimal point.**

   When you convert a decimal to a fraction, the answer is either a fraction (in which case your number has no digits to the left of the decimal point) or a whole number followed by a fraction, such as 2 ½ – the number to the left of the decimal point is simply the whole number.

2. **Count the number of digits to the right of the decimal point.**

   The bottom of the fraction is 1 followed by this many zeros.

3. **Work out the top of the fraction.**

   Write down all of the digits to the right of the decimal point. You can leave off any zeros at the beginning, but not zeros in the middle or at the end.

4. **Write down the top and bottom of the fraction.**

   Cancel down if possible (I explain cancelling in detail in Chapter 6).

5. **On a new line, write the whole number, if there's one, from Step 1, followed by the cancelled-down fraction.**

   You're done!

For example, here's what you do to convert 1.75 into a fraction:

1. **Leave the 1 alone.**

2. **You count two digits to the right of the decimal point.**

   So the bottom of the fraction is 100.

3. **The top of the fraction is 75.**

4. **The fraction is $^{75}/_{100}$.**

   You can cancel down to $^{15}/_{20}$ and then $^{3}/_{4}$.

5. **Bring back the 1 from earlier and write it in front of the fraction.**

   Your answer is 1 $^{3}/_{4}$.

Converting fractions to decimals

A fraction is really a divide sum – you can think of the bar in a fraction as a divide sign. So if you want to convert a fraction to a decimal, you just divide the top number by the bottom number – have a look at Chapter 4 on dividing if you could use a refresher.

When you divide decimals, you have to put a dot at the end of the number you're dividing, the one that goes 'under the bus stop' when you write it out. You then add the right number of zeros afterwards. Remember: sticking zeros after the decimal point is like adding no 10p pieces or no pennies – you don't actually change the number but you write a number that can then split up evenly if you want.

Follow these steps if you want to convert a fraction into a decimal number:

1. **If your fraction has a whole number in front, ignore that number for the moment.**

   You bring the whole number back in at the end.

2. **Draw a 'bus stop', as I explain in Chapter 4 on dividing.**

3. **Under the bus stop, write the number you want to divide, followed by a decimal point and as many zeros as you think you need.**

   Three zeros are probably plenty, but leave space at the end of the bus stop in case you need more – and there's no harm in having too many zeros!

4. **Put a dot above the bus stop, directly above the decimal point in the number below.**

5. **In front of the bus stop, write the number you want to divide by.**

6. **Divide as usual.**

   Completely ignore the decimal point.

7. **If you get to the end of the number with a remainder left over, add a zero to the end.**

   Carry on until you have no remainder.

8. **If you had a whole number in Step 1, add the whole number back in just before the decimal point.**

   For example, if your answer so far is 0.75 and you ignored a 1 in Step 1, your final answer is 1.75.

In Figure 7-2 I show you how to convert 4 ⅜ into a decimal.

Ignore the 4 and do 3 ÷ 8

Figure 7-2:
A worked
example of
converting a
fraction into
a decimal
number.

$$8\,\overline{)\,3\,.\,0\,\,0}$$

I put another dot above the first one so I don't forget it later. I'm going to ignore it otherwise.

$$\begin{array}{r} 0\,.\,3\,7 \\ 8\,\overline{)\,3\,.\,{}^30\,{}^60\,{}^4} \end{array}$$

I've run out of numbers but still have a remainder – need to add another zero.

$$\begin{array}{r} 0\,.\,3\,7\,5 \\ 8\,\overline{)\,3\,.\,{}^30\,{}^60\,{}^40} \end{array}$$

Now I have no remainder, so the divide bit is done!

I should remember to add the 4 from the beginning – my final answer is 4.375.

Remembering some common fractions and decimals

Certain fractions and decimals come up over and over again, so I list them in Table 7-1. If you get bored of working them out each time, try learning the numbers in the table.

Table 7-1	Common Fractions and Decimals	
Decimal	*Fraction*	*Comment*
0.01	$\frac{1}{100}$	£0.01 is a penny – and 100p is £1.
0.05	$\frac{1}{20}$	Twenty 5p pieces make £1.
0.1 (or 0.10)	$\frac{1}{10}$	Ten 10p pieces make £1.
0.2 (or 0.20)	$\frac{1}{5}$	Five 20p pieces make £1.
0.25	$\frac{1}{4}$	In the USA, a quarter is 25¢.
0.5 (or 0.50)	$\frac{1}{2}$	Two 50p pieces make £1.
0.75	$\frac{3}{4}$	Three times as big as a quarter.

Doing Sums with Decimals

Adding and taking away decimals follows exactly the same process as adding and taking away normal numbers. The only difference is the numbers have dots in and you have to line up the dots before you do the sum.

The most common problem my students have with doing decimal sums is not setting out their sums neatly and getting confused about what to add to what. Don't be like them. Setting your sums out in neat columns, especially when you work with decimals, saves you more headaches and time than you can possibly imagine.

In this section I show you how to add, take away, times and multiply with decimals. I also give you a few pointers about doing decimal sums by thinking about money or measurements.

Adding and subtracting: A quick review

Here I remind you quickly how to add and take away with normal numbers. If you feel comfortable with adding and subtracting normal numbers, feel free to skip this section and jump straight to 'Dealing with the dot'.

To add two numbers you follow these steps:

1. **Write the two numbers in neat columns, one above the other.**

   Make sure the numbers end in the same place and give yourself plenty of space between the columns.

2. **Starting at the right-hand end, add up each column.**

   If the sum of a column is less than ten, write the answer below the column. If the answer for a column is more than ten, write the last digit of the answer under the column and put an extra 1 at the bottom of the next column to the left.

3. **Keep going until you've done all the columns.**

   The numbers you've written down form your answer.

To take one number away from another you follow this recipe:

1. **Write the bigger number above the smaller number.**

   Make sure you use neatly spaced columns and the columns end in the same place.

2. **Starting from the right-hand column, see if the upper number is bigger than the lower number.**

   If the upper number is bigger, or the two numbers are the same size, jump to Step 4.

3. **If the top number is smaller, you need to borrow from another column.**

   Take one away from the next column to the left and add ten to the column you're working on. If the next column is zero, that column needs to borrow from its neighbour, and so on.

4. **Take away the bottom number from the top number.**

   Write the answer underneath the column.

5. **Repeat this until you run out of columns.**

   The numbers you've written down form your answer.

Dealing with the dot

The only difference between doing an add or take-away sum with decimals compared with a whole number is how you line up the numbers.

With decimal numbers, instead of lining up the ends of the numbers, you line up the dots. The dot marks the end of the whole number, so by lining up the dot you add units to units, tens to tens, and so on.

If your numbers are different lengths, either before or after the dot, the easiest way to avoid confusion is to fill in the gaps with zeros. For example, if you have to work out 29.993 – 1.23, you'd put a zero before the 1 (so both numbers have the same number of digits, in this case two, before the dot) and a zero after the 3 (so both numbers have the same number of digits, three, after the dot). You'd write that out as:

$$29.993$$
$$- \quad 01.230$$
$$= \quad 28.763$$

In Figure 7-3 I show you two examples of adding and taking away with decimals.

Figure 7-3:
Adding and taking away decimals.

```
  1 2 . 7 3          ͥ1ͥ2 . 7 3
+ 0 6 . 7 0        – 0 6 . 7 0
  1 9 . 4 3              6 . 0 3
        1
```

If your decimal sums have just two decimal places, try thinking about the numbers as money. You can work on the 'pence' – the numbers after the dot – and the pounds – the numbers before the dot – separately. You can even draw pictures of the coins and notes if you find that easier.

Unfortunately, the money way is somewhat limited. If you have more than two numbers after the dot, you can't think in pence any more.

Multiplying and dividing with decimals

Timesing and dividing with decimal numbers – at least as far as the numeracy curriculum goes – is no more difficult than doing the same thing with whole numbers. You perform the sums exactly the same way as before. The only difference is you need to put the dot in the right place at the end.

For the numeracy curriculum, you only need to times and divide one decimal number by a whole number. As you go further in maths, you may need to times and divide numbers that both have decimal points in – harder, but not overly complicated if you can estimate answers well.

Working out decimal times and divide sums follows the same process as the whole-number process I outline in Chapter 4. The devil, as usual, is in the details. I know two sensible ways for doing decimal sums: in one method you ignore the dot until the very end, and then put it back in. In the other method you work with the dot in place and try not to get distracted by it.

I prefer adding the dot at the end, but try both ways and see which works for you. You can turn a times or divide sum from a decimal monster into a nice whole number sum by temporarily ignoring the dot. To work out 53.2×8, you start by working out 532×8 and then put the dot back in the correct place. Similarly, to do $64.32 \div 4$, you work out $6432 \div 4$ and then add the dot. The trick is knowing where to put the dot.

The no-dot method I describe in this section works when only one of the numbers has a decimal point. If both numbers are decimal – which is beyond the scope of this book – you use a slightly different method to find where the dot goes at the end.

Here's my recipe for timesing a decimal number (I give an example in Figure 7-4 to help you):

1. **Work out the sum as normal.**

   Ignore the dot at this point.

2. **In the original number with a dot in it, count the number of digits between the dot and the end of the number.**

3. **In your answer to Step 1, count back the same number of spaces from the end of the number as your answer in Step 2 and put in a dot at that point.**

5.99×6 Estimate: $6 \times 6 = 36$

	500	90	9
6	3000	540	54

$$\begin{array}{r} 3000 \\ 540 \\ +\ \ 54 \\ \hline 3594 \end{array}$$

Figure 7-4: A decimal $\longrightarrow$ 2 dp in 5.99, so 2 dp in answer
times sum.

Answer   35.94

Always estimate your answer first so you know if you're in the right ballpark. Decimal divide sums are even easier: you work out the sum as normal and simply copy the dot directly up so it stays in the same place. I show you what I mean in Figure 7-5, where I work out 7.92 ÷ 6.

Figure 7-5:
A decimal divide sum. Left: Remember how to do 792 ÷ 6 . . . Right: This sum is identical, except you have a dot.

$$\begin{array}{r} 1 \quad 3 \quad 2 \\ 6\overline{)7 \ {}^{1}9 \ {}^{1}2} \end{array} \qquad \begin{array}{r} 1 \ . \ 3 \quad 2 \\ 6\overline{)7 \ . \ {}^{1}9 \ {}^{1}2} \end{array}$$

Flipping into fractions, and vice versa

If you prefer sums with fractions than with decimals, you can convert your decimal sum into a fraction sum. I don't really recommend this as a general practice, because the more steps you add to a question, the more likely you are to make a mistake. That's as true for me as for you.

If you need to do a divide sum, be careful. I don't cover dividing by fractions in this book. If you already know how to divide by a fraction, great; if not, I suggest you stick with decimal dividing for now, unless you feel like looking up how to divide by a fraction online or in another book.

Equally, if you love decimals and have a nasty fraction sum, you can convert the sum into a decimal sum. The only thing you really need to look out for here is the dreaded *recurring decimal*, which goes on forever and tends to mess up your sums. If a recurring decimal crops up, bad luck – you need to do the sum with fractions. I cover recurring decimals in more detail in the next section.

Doing decimals with a calculator

In real life you usually have recourse to a calculator or computer. Remember, though: your calculator and computer are powerful machines, but if you use them carelessly they give you the wrong results.

Doing decimals on a basic calculator

You probably use a *basic* calculator. I don't mean the machine's no good – I just mean it doesn't have all the intimidating bells and whistles that you see on a scientific calculator. Basic calculators always give you a decimal answer when you type in a sum. Just watch out for the following blips:

- ✔ **Typing errors:** One weakness of the basic calculator is that it doesn't have a 'natural display' – you can't see the sum as you type it in. You can't check what you type, so you can't see whether you've hit a wrong key or missed out a zero.

- ✔ **Manual fractions:** Getting a basic calculator to accept fractions is surprisingly tricky. I give you a workaround for this later in this section.

- ✔ **Recurring decimals:** You sometimes get an answer that seems to go on forever, for example, if you divide by 3 or 7. I give you a workaround for this later in this section.

Here's my fail-safe way of coping with recurring decimals:

1. **Work out the fraction part of the first number in the sum.**

   Type in the top number, then the ÷ key, and then the bottom number.

2. **If you have a mixed fraction (a whole number plus a decimal), press the + key, type the whole number, and then press the = sign.**

   Write down the answer. If the number is a recurring decimal or goes on for a while, round it to the first few decimal places – four is usually plenty.

3. **Repeat Steps 1 and 2 for the second number.**

4. **Type in the decimal number you wrote down in Step 2.**

   Then press the operation you want to do (+, −, ÷ or ×) followed by the number you wrote down in Step 3.

5. **Press the = key.**

   The calculator shows your answer.

This method only gives you a rough answer, but the answer is usually close enough for practical purposes.

Dealing with decimals on a scientific calculator

The good news is, most scientific calculators allow you to type in fractions and decimals in a simple, fairly obvious way. The bad news is they tend to display the answers in *surd form*, meaning you see fractions, square roots and the like all over the place – great if you're an A-level student, but frustrating if you want some basic decimals.

Throughout this section, I assume you're using the most common design of scientific calculator – one with a circular button just below the screen at the top. If you have a different kind of calculator, you can try to follow along as best you can, or you can work through the instructions for a basic calculator in the previous section; these will work just as well. Happily, scientific calculators have a button to change fractions into decimals. Look for the S↔D key, usually towards the bottom right of the small keys, just above the red buttons. The S stands for 'surd' and D for 'decimal'; the arrow thing means 'switch between'. Pressing the S↔D button simply switches from surd form – with all the ugly symbols – into nice, easy decimals.

Most scientific calculators let you type in fractions without having to mess about with adds, divides and writing things down. Look for the fraction button up towards the top left of the calculator. It usually looks like two boxes, one on top of the other, with a line between them.

Here's how to enter a fraction that isn't a mixed number (with no additional whole number):

1. **Press the fraction button.**

   You should see two boxes, one on top of the other, with a line between them. The top box contains a flashing line.

2. **Type the top of the fraction in the box with the flashing line.**

3. **Press the down arrow on the circular button just below the screen.**

   This moves the cursor to the bottom box.

4. **Type the bottom of the fraction in the bottom box.**

5. **Press the right arrow on the circular button to move out of the fraction.**

   Now write the rest of your sum.

To type in a mixed number, say 1 ½, press shift before the fraction button. The calculator displays three boxes, like those in Figure 7-6. Then just type in the numbers as before, and use the circular button to move between the boxes.

Figure 7-6:
The mixed-
fraction
display.

 If you mess up, press the red DEL button to delete what you just did, or the AC button to clear the whole screen and start again. Some older calculators use 'CE' to clear the last step and 'C' to clear everything.

Chapter 8

It's All Relative: Ratios and Proportions

Ratio and proportion are both about the relative size of things. If I say I'm going to split something with you 50:50, I'm using a ratio to describe how big your share is in comparison with mine – in this case, the two shares are the same. If we split something 60:40, one of us has a bigger share than the other – the ratio tells us how much bigger.

Proportion is similar to ratio, only it generally says something like 'you've got twice as much as I have' or 'I need three times as many of all the ingredients.'

The difference between proportion and ratio is really all in the way you write them: ratios have a colon between two (or more) numbers, and proportion says directly 'This is so many times bigger or smaller.'

Since proportion and ratio are so similar, I tackle them both the same way, using a tool called the Table of Joy, which I lay out for you in this chapter. The Table of Joy is an insanely useful tool. I use it in all sorts of chapters in this book to help you work out percentages, draw pie charts and any number of other things.

Don't worry if the Table of Joy isn't your thing. I also show you another way of dealing with ratios and proportions.

Meeting the Table of Joy

The Table of Joy is an incredibly useful method for working out any kind of problem involving proportions – that is to say, if you double one of the numbers, the other automatically doubles. For example, the price of most things is proportional to how much you buy; two tins of beans cost twice as much as one tin of beans does. In the same way, twelve tins of beans cost twice as much as six tins.

Among other things, the Table of Joy in its basic form can be used for the following:

- **Currency conversion:** If you need the exchange rate, or want to know how many pounds you get for a number of euros, turn to the Table of Joy.

- **Percentages:** Whether you want to find one number as a percentage of another, the percentage increase or decrease, or an amount with or without tax . . . the Table of Joy is your friend.

- **Pie charts:** If you want to know the angle, the value of a slice or the total value of a pie chart, you can use an easy sum . . . the Table of Joy reminds you of that sum.

- **Proportion:** If you have two linked values where doubling or tripling one automatically triples the other (if I drive three times as far, I use three times as much petrol), the values are *proportional*. And the Table of Joy deals with that.

- **Ratios:** Whenever you see a colon, crack your knuckles and say 'Here's a Table of Joy question!'

- **Scaled maps and drawings:** How far is that in real life? How far is that on the map? What's the scale? Ask the Table of Joy and it will answer you truly.

- **Speed–distance–time calculations:** Given any two of these three, you can work out the other using the Table of Joy.

- **Unit conversion:** Apart from temperatures, which are a bit squiffy, you can use the Table of Joy to do just about any unit conversion in this book.

Glancing through an old numeracy test, I reckon you can use the Table of Joy in the answers of somewhere between a third and a half of the questions. I can't stress too much how useful this little baby is, both for numeracy and for further studies – even at GCSE I see exam papers where you can use the Table of Joy for around half of the available marks.

The story of the Table of Joy

The Table of Joy is based on a way of working out similar sums called the Rule of Three, but I find this name is less obvious and less delightful than the Table of Joy. For a long while, the Table of Joy was just called The Table. Then one of my students started using it and suddenly started answering dozens of questions correctly that she hadn't had a clue about before. She decided it ought to be called the 'Table of Doom'. Then she thought for a moment and changed her mind. 'Not doom! That makes it sound like a bad thing. It should be the Table of Joy.'

Good name, I thought. Good name.

Introducing the Table of Joy

The Table of Joy – as its name suggests – is a table. You use the table to lay out the information you already have so you know precisely which sum you need to work out. The table doesn't solve the sum for you – you still need to do that yourself – but the table takes all the guesswork out of deciding what to times and divide by what.

To create a Table of Joy, you draw a three-by-three grid like you play noughts and crosses on, label the rows and columns, decide where to put three numbers that you either have already or can easily work out, and then do a sum that has the same structure each time.

One of the many beauties of the Table of Joy is that it doesn't matter which way around you label the rows and columns, as long as you put the numbers in a sensible place. The sum is exactly the same if you swap both of the rows with the columns or if you swap the two columns or the two rows over. You can't swap a row with a column, though.

Seeing how the Table of Joy works

There are five steps to using the Table of Joy to solve a problem:

1. **Draw a fairly big noughts-and-crosses grid.**

   It needs to be big enough so you can write labels in the rows and columns.

2. **Label the rows and columns with the names of relevant information from the question.**

3. **Fill in the numbers relevant to each row and column, and a question mark for the square representing the answer you want.**

4. Circle the number in the same column as the question mark and the number in the same row as the question mark.

Write these numbers with a times sign between them. Then write a divide sign followed by the remaining number.

5. Work out the answer!

The main trick of using the Table of Joy is working out what labels to use and which number to put where.

In Figure 8-1 I show a Table of Joy in various stages of completion, solving the following question:

Larry and Curly split the loot in the ratio 3:7. Curly walks away with $350. How much loot does Larry receive?

	Larry	Curly
Ratio		
Loot		

Figure 8-1:
The Table of Joy being used to answer a very tricky question.

	Larry	Curly
Ratio	3	7
Loot		350

$$\frac{350 \times 3}{7} = 150$$

Understanding what goes where

Labelling the rows and columns is a very important part of the Table of Joy. This labelling may seem like a chore, but it really helps you to keep track of which number goes where.

The way I normally do my labelling – and I invented the Table of Joy, so I get to choose – is to write the things I'm counting or measuring across the top. Down the side, I put the different pieces of information I have or want.

For the example in Figure 8-1, the things I am counting are Larry's money and Curly's money. Down the side, I write the ratio and a question mark, because I want to know the total amount of money.

Labelling in this way makes the meaning of each number much more obvious. If you've just got a noughts-and-crosses grid, figuring out where to put Curly's part of the ratio is really hard – but with the label 'Curly' at the top and 'ratio' at the side, things are more obvious.

Getting Rational: Understanding Ratios

A ratio describes how many times bigger one share of something is than another share. If you say 'We'll split the money 50:50', you mean that, however much money there is, you get half and I get half. If we agree on a 2:1 split, you get twice as much as I do.

The ratio tells you about the *relative* sizes of the shares. You can't tell from the ratio whether we split £3 or £300 million, only that you received twice as much as I did. (You're welcome to it. It's only pretend money.)

You meet ratios when you mix things – for example, cocktails or concrete – and when you try to divide things between several people. The traditional examples for splitting things up are money, people and sweets. I have no idea why sweets are so important in the scheme of things.

You also see ratios in the scale on maps. I talk more about that, and maps in general, in Chapter 14.

One for you, two for me: Sharing

Picture the cinema cliché of bad guys sitting around a table and splitting up their ill-gotten gains: 'One for you, two for me. One for you, two for me . . .' These chaps are dividing their money in a ratio of 1:2. For every dollar the other person gets, the first person gets two.

If they start with three dollars, the other guy would get one dollar and the first would get two. If they start with 30 dollars, the split would be 10 and 20. The key thing is that however much money is in the pile to begin with, the first guy gets twice as much as the other.

Cancelling ratios

Exam questions often ask you to 'reduce a ratio to its simplest form'. The reason for this is to make the numbers as small as possible – it's much easier to see the relative size of two things if their ratio is written as 1:4 than, say, 90:360. It also makes the Table of Joy sums easier.

Reducing a ratio to its simplest form is very similar to *cancelling down* a fraction to its simplest form, which I cover in scary detail in Chapter 6 – fractions and ratios are very closely related. It's ok, though, you don't need to be a fractions expert to work with ratios!

Here's the recipe for cancelling ratios:

1. **Find a number that you can divide both sides of the ratio by so you still end up with a whole number.** Sometimes there's no such number – if that's the case, the ratio is already in its simplest form.

2. **Divide each side of the ratio by that number.**

3. **Go back to Step 1 until you can't find any more numbers to divide by.**

4. **Write down the numbers in the same order as in the original ratio, with the colon between them.**

In maths exams, more often than not you can cancel ratios all the way until one of the sides of the ratio is 1. Sometimes, though, the exam asks you to cancel until one of the sides of the ratio is 1, even if the other side isn't a whole number. (You may see the question 'Write the ratio in the form n:1' or 'Use the form 1:n.') To do this, divide the bigger number by the smaller number and write down the ratio in the right order – the bigger number should always be on the same side as the bigger number in the original ratio.

For example, you can cancel 60:40 to 3:2 (in whole numbers) or 1.5:1 (in the form n:1).

Applying the Table of Joy

A typical ratio question might tell you that someone splits their time between Project A and Project B in a ratio of 5:3. If he spends 15 hours this week on Project B, how long does he spend on Project A?

In this question, you have three key pieces of information: the two sides of the ratio (5 and 3) and the time for Project B (15). You can put these numbers into the Table of Joy. Before I outline the idea in detail, see if you can figure out what goes where.

Here's what you do (I also show the steps in Figure 8-2):

1. **Draw a noughts-and-crosses grid.**

   Leave plenty of space for the labels.

2. **Figure out what the two sides of the ratio represent.**

   In our example, the two sides of the ratio represent Project A and Project B. These labels go in the top row.

3. **On the left, write 'ratio' in one row and whatever you want to measure in the other – in this case, 'time'.**

4. **Fill in the ratio in the middle row (remember to keep it the right way around) and put the time you know in the correct column.**

   Here, the time goes in the column marked 'Project B'. Put a question mark in the remaining square.

5. **Write out the Table of Joy sum.**

   Take the number in the same column as the question mark, times by the number in the same row as the question mark, divide by the number opposite. For our example, you do $5 \times 15 \div 3$.

6. **Work out the sum.**

   In our example, the answer is 25.

Figure 8-2:
Using the Table of Joy to solve a ratio problem.

	Project A	Project B
Ratio	5	3
Hours	?	15

$$\frac{5 \times 15}{3} = 25$$

You'd be a little bit lucky to get a question as simple as the one above in an exam. More often than not, you have to work with the total and one side of the ratio.

The good news is this isn't any more difficult – you simply need an extra layer of thought before you do the Table of Joy sum.

Take a moment to read the question and decide whether you're interested in the two parts or a part and a total. The example above looks at the two-parts version.

By contrast, with the part-and-a-total version, you replace the missing part of the ratio with the total of the ratio – just add the numbers either side of the colon.

After that, you follow exactly the same process, but change one of the column labels to 'total' and the numbers in that column as appropriate. I give an example in Figure 8-3 to answer the following question:

Alice and Bob share 12 biscuits in the ratio 3:1. How many does Alice eat?

Figure 8-3:
Ratios, totals and the Table of Joy.

	Alice	Total
Ratio	3	4
Biscuits	?	12

$$\frac{3 \times 12}{4} = 9$$

Working with parts

If you don't like the Table of Joy, the more traditional way of doing ratios is to think about the value of a share. The ratio 2:7 means that for every two shares in the first group, there are seven shares in the other. If you divide some biscuits in the ratio 2:7 and the first group gets 50, each share is worth 25 biscuits. From there, you can say that the other group has seven shares, or $7 \times 25 = 175$ biscuits, and there are nine shares altogether: $9 \times 25 = 225$.

Here's the recipe for working out ratios using the traditional method:

1. **Work out how many shares the number you have represents.**

   You need to add up the parts if you know the total.

2. **Divide the number you know by the answer in Step 1.**

   This is the value of each share.

3. **Times your answer to Step 2 by the number of shares you're interested in.**

   This is your answer.

The method I describe above has fewer steps than the Table of Joy method, but the Table of Joy makes obvious which sum to do. You're the boss, so use whichever method works best for you.

Managing multiple ratios

Sometimes you have three or even more numbers in a ratio. You might see a drink made of orange juice, apple juice and pineapple juice mixed in the ratio 3:1:1 – which means that for every three measures of orange juice, you add one measure of apple juice and one measure of pineapple juice.

To deal with multiple ratios, you use the same method as for ratios with only two numbers. You can work with pairs of numbers from the ratio or a number and the total using the Table of Joy, or you can work out the value of a share as before. Here's an example question:

> *You want to make a litre of drink containing orange juice, apple juice and pineapple juice in the ratio 3:1:1. How much of each juice do you need?*

To answer this question using the share method, follow these steps:

1. **There are 1,000 millilitres in a litre and the ratio adds up to five, so each share is worth 1,000 ÷ 5 = 200ml.**

2. **Three shares of orange juice is 3 × 200 = 600ml.**

3. **One share of apple juice is 1 × 200 = 200ml.**

4. **One share of pineapple juice is 1 × 200 = 200ml.**

In Figure 8-4 I show you how to reach the same conclusion with the Table of Joy.

Figure 8-4:
A multiple ratio. Left: the sum for orange juice. Right: the sum for apple juice. The sum for pineapple juice is the same as the one for apple juice.

	Juice	Mixture
Ratio	3	5
Total	?	1000

$$\frac{3 \times 1000}{5} = 600$$

	Juice	Mixture
Ratio	1	5
Total	?	1000

$$\frac{1 \times 1000}{5} = 200$$

Getting a Sense of Proportion

Two quantities are *proportional* if, whenever you double one, the other also doubles. For example, speed and distance are proportional: if you drive for an hour at 60 miles per hour, you travel twice as far as you do if you drive for an hour at 30 miles per hour. An example that often comes up in maths exams concerns the amount of each ingredient you need for a recipe. The recipe is proportional to the number of people the dish feeds. If you double all of the ingredients, you can feed twice as many people with the end result.

Proportion doesn't stick at doubling. If you know two quantities are proportional, you can multiply them both by the same number and end up with the appropriate relationship. For example, if you drive at 20 miles per hour, you get a third as far as if you drive at 60. Likewise, if you multiply your recipe ingredients by seven, you can feed seven times as many people.

The reason the Table of Joy works in so many topics is that quite a lot of maths – particularly in the numeracy curriculum and at GCSE – is based on proportional quantities, and that's exactly what the Table of Joy handles best.

In maths exams, proportion questions generally involve taking two proportional values and *scaling* them – finding out what happens to one of them when the other changes. The good news is you can use the Table of Joy to find the answer. The other good news is that the Table of Joy isn't the only way to answer these questions – although I think the other options are trickier.

Typical proportion questions in maths tests involve adjusting a recipe so you can feed more people, or enlarging the measurements of a shape.

Defining proportion

Having a sense of proportion is supposed to be a good thing. Artists spend years studying so they can draw the proportions of the human body correctly. But what exactly is proportion?

A decent definition of proportion is this:

> *The proportion between two values is how many times bigger one is than the other.*

To explain that two things are proportional, you would normally say something like 'miles and kilometres are proportional in the ratio 1:1.6' or '1 mile is the same as 1.6 kilometres'. The number of miles is a fixed multiple of the number of kilometres, so the two quantities are proportional.

Perfecting proportions with the Table of Joy

Here's how you answer a proportions question using the Table of Joy:

1. **Draw a noughts-and-crosses grid.**

   Leave plenty of space for labels.

2. **Label the columns to reflect what's changing.**

   For example, use 'small' and 'large', or 'before' and 'after'.

3. **Label the rows with the information you're interested in.**

   For example, you may use 'people' and 'grams' if you're doing a recipe, or 'short side' and 'long side' if you're working on a shape.

4. **Fill in the numbers you already know in the appropriately labelled cells.**

   Put a question mark in the empty cell.

5. **Write down the Table of Joy sum.**

   Times the number in the same column as the question mark by the number in the same row as the question mark, and divide by the one opposite.

6. **Work out the sum.**

   That's your answer.

In Figure 8-5 I show a worked example for the following question:

One mile is about 1.6 kilometres. How many kilometres is 20 miles?

Figure 8-5:
A proportions example.

	miles	km
conversion	1	1.6
distance	20	?

$$\frac{20 \times 1.6}{1} = 32$$

Another way to work out proportions is to use the 'value of a share' approach that I use for ratios in the section 'Working with parts' earlier in this chapter. (In fact, ratios and proportions are so similar that treating them as different things is almost fraud.)

For example, with a recipe, you work out how much of an ingredient you need for one person and then times that amount by the number of people you want to feed. Here's what to do:

1. **Divide the amount of ingredients you're interested in by the number of people the recipe is for.**

2. **Times this by the new number of people you want to feed.**

You probably don't think this two-step process looks more complicated than the six-step Table-of-Joy process I describe earlier. But try the two-step process with the example in Figure 8-5 . . . you get a pretty unpleasant fraction when you do the divide in Step 1, something the Table of Joy avoids.

Applying proportion

One of the reasons maths is such an important subject (and, to me, such a beautiful one) is that the same methods often apply across a broad range of subjects. You can apply proportion in dozens upon dozens of real-life situations (and, of course, many different questions in an exam). Here are some examples of times when you may need to apply proportion:

- **Capacities:** If you know how many glasses a bottle of water fills, how many glasses would seven bottles fill?

- **Coverage:** If a tin of paint covers $4m^2$ of wall, how many tins do you need to paint a room with wall space of $48m^2$?

- **Fuel usage and cost:** If your car uses one litre of petrol to travel 11 miles, how far can you travel on a full 45-litre tank?

- **Prices:** What is the total cost or the cost per item? How many items can you buy with a certain amount of money?

- **Scaling recipes:** If you have a recipe for four people, how much food do you need to feed ten people?

- **Typing speeds:** If you can type a page of drivel . . . I mean, novel, at 50 words per minute and work solidly for an hour, how much novel do you have at the end of your hour?

- **Umpteen others:** Anything where doubling one value doubles the other can lead to a question where a proportion sum is appropriate.

Scaling recipes

One kind of proportion question seems to come up more often than any other in maths exams, and it involves scaling recipes. Imagine you have a recipe that serves a certain number of people, but you need to cook for a bigger or smaller party. The amount of each ingredient you need is proportional to the number of people it serves, so you need to adjust the quantities to account for your head count.

A typical maths exam question looks something like this:

> *Here are the ingredients for a recipe for a cake that serves four people:*
>
> *100g butter*
>
> *100g self-raising flour*
>
> *150g caster sugar*
>
> *2 eggs*
>
> *You want to make a cake that serves ten people. How much caster sugar do you need?*

You can solve this question using the Table of Joy. If you follow the instructions in the section 'Perfecting proportions with the Table of Joy', you get a table and sum that look like that in Figure 8-6.

Figure 8-6:
Solving a recipe problem with the Table of Joy.

	recipe	real life
people	4	10
sugar	150	?

$$\frac{150 \times 10}{4} = 375$$

If you're comfortable with decimals (which I cover in Chapter 7), you can work out that one portion would be 37.5 grams, so ten portions would require 375 grams.

Here's another example, now you're in the swing of things:

I just bought a 10 kilogram bag of cement, which has this recipe on the side of it: 'To make 100 kilograms of concrete, mix 10 kilograms of cement with 75 kilograms of aggregate, and 15 litres of water.'

Since this bag of concrete exists only in my imagination, I suggest you don't follow the recipe. Frankly, it's a toss-up as to whether this concrete recipe will give you something more or less edible than the cake recipe from earlier, and whether the cake or the concrete would be harder.

Unfortunately, I don't need 100 kilograms of concrete. I need 20 kilograms for the equally imaginary swimming pool I want to build in my dream mansion (which is, of course, only a dream).

I don't want to make 100 kilograms of cement, because I don't want to throw most of it away (even imaginary concrete is hard to get rid of). Instead, I want to scale the recipe down. Looking at the aggregate (the other ingredients work just the same way), I use the Table of Joy, as in Figure 8-7.

Figure 8-7:
Solving the
concrete
problem
with the
Table of
Joy.

	aggregate	concrete
recipe	75	100
pool	?	20

$$\frac{75 \times 20}{100} = 15$$

I need $(75 \times 20 \div 100)$ kilograms of aggregate, which works out to 15 kilograms. In the same way, I find that I need 2 kilograms of cement and 3 litres of water.

Another way of doing this sum is to work out the *scale factor* – how many times more concrete am I making? In this case, I'm making a fifth as much concrete as the recipe is for, so I need to find a fifth of each ingredient.

A fifth of 75 is 15 – so I need 15 kilograms of cement, just like before. You can check the other amounts too, if you like.

This method has two possible problems: the scale factor may not be obvious, and you may end up with an especially ugly fraction sum.

Chapter 9

Perfect Percentages, 100% of the Time

*I*f you listen to the news, particularly the financial stuff, you hear percentages bandied about all over the place. A share price may drop 5 per cent. Inflation may be 3 per cent. Train tickets may be 7 per cent more expensive. The percentage of adults with good numeracy skills may be on the increase.

For the numeracy curriculum, you need to deal with percentages written down rather than on a news broadcast. If you've nailed Chapters 6 and 7 on fractions and decimals, I think you may find this chapter on percentages quite easy. If you haven't read those chapters yet, I suggest you work through them before you start looking at percentages.

Per cent simply means 'for every hundred'. We often write per cent as a squiggle like this: %. The symbol looks a little like a divide sign rotated slightly. Per cent has exactly the same meaning as 'divided by 100', so 23% is the same thing as $^{23}/_{100}$ and 0.23.

Perusing Some Percentages You Already Know

You can hardly avoid seeing percentages at the bank and in the shops. Banks give interest rates on savings accounts, loans and mortgages as percentages. You may have a savings account with an interest rate of 5%. What this (roughly) means is that every year, the bank looks at your balance, works out 5% (or $\frac{5}{100}$) of that balance, and adds that much money to your account. Loans work the other way around. If you have a loan on which you pay 10% interest, each year the bank looks at how much you owe, works out 10% of that sum, and says 'You owe us this much more now.'

Modern banks don't work quite like this. Instead they do a calculation every day and update your account each month, but the difference in the amount you accrue or pay is negligible.

You also see percentages in the shops, often in the sales. You may see signs screaming '30% off!' (normally with 'up to' in very small print). This means the shop has taken the original price, worked out 30% ($\frac{30}{100}$ or $\frac{3}{10}$) of that amount, and then taken that much money off the price.

Comparing Percentages, Decimals and Fractions

Ten per cent is the same thing as $\frac{1}{10}$, and 25 per cent equals 0.25. You may well wonder why we use so many different ways of saying the same thing. This isn't an easy question to answer. The best answer I have is that in different contexts you need to communicate in different ways – just like you use one language to persuade your five-year old nephew to stop playing with his noisy toys and another language to persuade your sister not to bring her son's noisy toys next time.

For example, saying 'Prices went up by 5%' is easier than saying 'Prices increased by a factor of 1.05' or 'Prices went up by a twentieth.' Having said that, we can use decimals, fractions and percentages to express the same thing. In this section I show you how to convert percentages into decimals and fractions, and vice versa.

Percentages and decimals

Converting a percentage to a decimal is easy – just remember 'per cent' means the same as 'hundredths' or 'divided by 100'. Here's what you do to convert 25% to a decimal:

1. **Write down the number of per cent.**

   In our example you write 25.

2. **If the number doesn't have a decimal point already, put one at the end.**

   Now you have '25.'.

3. **If the number is less than 100, put a zero at the beginning.**

   In our example you have '025.'.

4. **If the number is less than 10, put another zero at the beginning.**

   Twenty-five is bigger than 10, so we miss this step out.

5. **Move the decimal point two places to the left.**

   In our example you get 0.25.

6. **If the number has zeros at the end, ignore them.**

As another example, 10% becomes 010. and then 0.10. You ignore the zero at the end and write '10% = 0.1'.

Converting a percentage to a decimal is almost exactly the same as converting a price in pence into pounds: 25p is the same as £0.25. The only difference is you don't ignore a zero at the end in money sums, so you write 10p as £0.10 rather than £0.1.

To convert a decimal into a percentage, you work in the opposite way. Here are the steps using 0.4 as an example:

1. **Write down the decimal.**

   In our example you write 0.4.

2. **If you have only one digit after the dot, write a 0 on the end.**

   You write 0.40.

3. **Move the decimal point two places to the right.**

   You now have 040.

4. **If you have any zeros at the beginning of the number, before the decimal point, ignore them.**

   You write 40.

5. **Write a % at the end. If the decimal point is at the end of the number, you can ignore it, too.**

   Your answer is 40%.

Percentages and fractions

Converting a percentage to a fraction is as easy as cancelling. Here's my easy recipe:

1. **Write down the percentage.**

   In place of the % sign, write '/100'. You now have a fraction – for example, $25\% = {}^{25}\!/_{100}$.

2. **Find a number you can divide the top and the bottom by.**

   Good numbers to look for are 2, 5 and 10. If you can't find any number to divide both the top and the bottom by, you're done. In our example you can divide top and bottom by 5.

3. **Divide the top and bottom by the number in Step 2.**

   In this case you'd get ${}^{5}\!/_{20}$.

4. **Repeat Steps 2 and 3 until you're done.**

   You can divide the top and bottom numbers by 5 to get ¼. So $25\% = ¼$.

Going from a fraction to a percentage is a little harder, but the Table of Joy is a good friend to you here. Get the low-down on the Table of Joy in Chapter 8 if you don't know how to use it. Here's my fractions-to-percentages recipe using the Table of Joy:

1. **Draw a noughts-and-crosses grid.**

2. **Label the middle column 'fraction' and the right column 'per cent'.**

3. **Label the middle row 'top' and the bottom row 'bottom'.**

4. **Fill in the top and bottom of the fraction in the middle column.**

5. **Write 100 in the bottom-right square (remember 'per cent' means 'over 100').**

   Put a question mark above the square containing 100. You need to find the number in the question-mark square.

6. Do the Table of Joy sum.

Times the question mark's neighbours together (the top of the fraction times 100) and divide by the opposite number (the bottom of the fraction). The number that comes out is your answer.

Have a look at Figure 9-1, where I show the Table of Joy in action converting a fraction to a percentage.

Figure 9-1:
Converting a fraction into a percentage with the Table of Joy.

	fraction	percent
top	1	
bottom	4	100

$$\frac{1 \times 100}{4} = 25$$

Working Out Percentages the Traditional Way

At school, your maths teacher probably spent hour after tedious hour trying to persuade you that working percentages out with fractions and multipliers and so on was (a) simple and (b) fun. The poor delusional fool. I only go through this method here in case you understood it and need a reminder. Personally, I prefer the Table-of-Joy method, which I explain in the section 'Working Out Percentages Using the Table of Joy' later in this chapter.

Percentages as hundredths

'Per cent' means just the same as 'hundredths', so you can work with percentages just as you can with fractions. If you know how to work with a few simple fractions, you can add and subtract percentages as you need.

The easy ones: 1%, 10% and 50%

One per cent – 1% – is the same thing as $\frac{1}{100}$. To work out 1% of any number, you just divide the number by 100. So, 1% of £2,000 is £20. If 40,000 people live in your town and 1% of them like percentages, you are (I hope!) 1 of 400 percentage fanatics nearby. You should start a club.

Ten per cent is the same as $\frac{10}{100}$ or $\frac{1}{10}$. To figure out 10% of a number, you just divide the number by 10. So if a school has 1,600 kids and 10% of them are left-handed, the school teaches 160 left-handed kids.

Fifty per cent is the same thing as a half. Fifty per cent is $\frac{50}{100} = \frac{1}{2}$. To work out 50% of a number, you just divide the number by two. If a person takes a test with 70 questions and needs 50% to pass, they have to answer 35 questions correctly to pass.

Combining easy percentages to make other percentages

If you can work out 1% of a number, you can work out any percentage with a simple times sum. Here's what you do:

1. **Work out 1% of the number and write it down.**

   You can easily work this out by dividing your number by 100. If you want to find 35% of 250, you work out that 1% of 250 is 2.5.

2. **Times your answer from Step 1 by the number of per cent you need.**

   You might want to take a refreshing look at Chapter 4 if you find multiplying tricky, and possibly Chapter 7 if you could use some help on decimals. In this example, you multiply 2.5 by 35 to get 87.5. This is your answer.

You can do something similar using 10% of a number. For example, imagine a company says 87% of the people it surveyed were satisfied customers. In the small print the company mentions that it actually surveyed 600 people. You work out how many people are happy as follows:

1. **Calculate 10% of your number and times by how many tens are in your percentage.**

   Here, 10% of 600 is 60, so 80% is $60 \times 8 = 480$.

2. **Calculate 1% of your number and times by how many ones are in your percentage.**

   Here, 1% of 600 is 6, so 7% is $6 \times 7 = 42$.

3. **Add the two numbers together to get your answer.**

   Here, you do $480 + 42 = 522$ which is your 87%.

Percentages of the whole

Another way to figure out a percentage of a number is to turn the percentage into a decimal and then multiply. This method is a bit ungainly, but it works if you're careful. Here are the steps:

1. **Convert the percentage into a decimal.**

   Look at the section 'Comparing Percentages, Decimals and Fractions' if you don't know how to do this.

2. **Times the original number by the decimal number in Step 1.**

3. **Write down your answer.**

This short recipe disguises a fairly nasty sum. For example, to work out 87% of 600, you do 600×0.87 – not nice! If you're confident with decimals, feel free to work this way – but if not I suggest you use the Table of Joy way that I describe in the section 'Working Out Percentages Using the Table of Joy'.

You may also want to know how many per cent something is. If you know that September has 30 days, and it rained on 21 of them, what percentage of September was rainy? The sum looks like this:

1. **Multiply the number that's not the whole thing by 100.**

   For our example, you do $21 \times 100 = 2,100$.

2. **Divide the answer to Step 1 by the 'whole thing' number.**

   For our example, $2,100 \div 30 = 70$.

3. **Write down the answer.**

   Seventy per cent of this hypothetical September was rainy.

You may need to approximate a percentage or find the nearest percentage from a list. In this case, you estimate the part and whole numbers by rounding them off appropriately before you do the percentage sum.

Some exam questions on percentages are very sneaky indeed. The question may, for instance, tell you the number of blue shirts sold and the number of white shirts sold and ask you to find what percentage of the total was blue. To do this, work out the 'whole thing' number before you start. Don't be tempted to just use the numbers in the question without thinking first. Here's an example question:

In a car park are 12 cars and 18 empty spaces. What percentage of the car park is full?

You start by working out the 'whole thing' number. You have 12 full spaces and 18 empty spaces, making a total of 30 spaces.

Twelve of the 30 spaces are full, so you work out $12 \times 100 = 1,200$, and then $1,200 \div 30 = 40$. So 40% of the car park is full.

Going up and going down

We commonly use percentages in the context of increases and decreases – for example, 'I got a 3 per cent pay rise' or 'The euro has lost 25 per cent of its value against the pound since Colin last went to France.' Increases and decreases add an extra step of complication to your percentage sums, but things are still straightforward if you keep your concentration.

Working out the difference first

If your water company plans to increase bills by 5% or your internet company is giving you a 10% discount, you may want to figure out what your new bill will be. In both instances, you do roughly the same thing:

1. **Work out the appropriate percentage of the whole bill.**

   For example, if your water bill was previously £40 and will increase by 5%, work out 5% of £40 = £2. Alternatively, if you pay £25 per month for internet and the company is giving you a 10% discount, you work out 10% of £25 = £2.50.

2. **Decide whether the change will make your answer bigger or smaller.**

   An increase means your number goes up. A decrease means your number goes down.

3. **If you have an increase, add your answer from Step 1 to the original amount.**

   For example, for your water bill you add £2 + £40 = £42.

4. **If you have a decrease, take away your answer from Step 1 from the original amount.**

   For example, for your internet bill, you do £25 – £2.50 = £22.50.

Working out the full percentages

Another way to deal with increases and decreases is to use the full-percentages method. Many students in my class at school got a bit flummoxed at this point. By which I mean, I remember finding this hard and yelling 'But I don't *understand*!' I was 13. It was sunny. I wanted to be outside playing football.

This method needs a bit of patience and thinking. The basic idea is to ask 'What percentage of the original number do I need?' and then calculate that percentage. You work out the difference like you did in the last section, but in a different order.

Here's what you do:

1. **Start from 100%.**

   Always always always.

2. **Decide whether you need to go up or down, and then add or take away that percentage.**

   For example, if you have a 5% rise, you end up with 105%. If you have a 7% drop, you end up with $100 - 7 = 93\%$.

3. **Work out that percentage of the original number.**

   For the examples in the previous section, you need to find 105% of £40 for the water bill, so you do $1.05 \times £40$. For the internet bill, you need 90% of £25, or $0.9 \times £25$.

Working Out Percentages Using the Table of Joy

The Table of Joy is perfect for anything that involves proportions. And percentages are one of those things. If you don't yet know the Table of Joy inside out, check out Chapter 8 before you read this section.

Percentages are proportional because the answer to '20% of something' is twice as much as the answer to '10% of the same thing' – that is, if you double the percentage, you double the answer.

Finding a percentage of the whole

To work out a number of per cent of a whole number with the Table of Joy is a piece of cake. Here's what you do:

1. **Draw a noughts-and-crosses grid.**

   Label the columns 'number' and 'per cent'. Label the rows 'part' and 'whole thing'.

2. **Put the number of per cent you want to find in the middle-right ('part'/'per cent') square.**

   100% is the whole thing, so write 100 in the bottom-right ('whole'/'per cent') square. You also know how many the whole thing is, so write that number in the bottom-middle ('whole'/'number') square.

3. **Put a question mark in the middle square ('part'/'number').**

   Its neighbours are to the right ('part'/'per cent') and below ('whole'/'number'). The square opposite is 100 ('whole'/'per cent').

4. **Write down the Table-of-Joy sum.**

   Times the neighbours together and divide by the opposite number.

5. **Work out the sum.**

For example, if a footballer scores 80% of his penalty kicks and took 15 such kicks last season, you can work out how many penalties he scored. Set up the Table of Joy as in Figure 9-2. The sum is $80 \times 15 \div 100 = 1{,}200 \div 100 = 12$. The footballer scored 12 penalties.

Figure 9-2:
Working out 80% of 15 using the Table of Joy.

	number	percent
part		80
whole	15	100

$$\frac{80 \times 15}{100} = 12$$

If you know two numbers, you can use the Table of Joy to work out what percentage one is of the other. Here's the recipe:

1. **Draw out the noughts-and-crosses grid.**

2. **Label the columns 'number' and 'per cent'.**

3. **Name the rows 'part' and 'whole'.**

4. **Fill in what you know.**

   The part number and the whole number go in the middle column, and you know the whole number corresponds to 100%, so that goes in the bottom-right square.

5. **Put a question mark in the remaining grid square, the middle-right.**

   Its neighbours are middle-middle and bottom-right. Opposite is bottom-middle.

6. Do the Table-of-Joy sum.

Times the neighbours together, and divide by the opposite number.

For example, if I own 12 mugs and 3 are clean (an unusually tidy day in my house), I fill out the Table of Joy as in Figure 9-3. I work out $3 \times 100 = 300$, and then divide that by 12 to get 25. Three mugs are 25% of 12 – or a quarter.

Figure 9-3:		number	percent
Working out 3 as a	part	3	
percentage of 12.	whole	12	100

$$\frac{3 \times 100}{12} = 25$$

Be very careful of questions that sneakily fail to give you the total. I could rewrite the previous example slightly and tell you I have three clean and nine filthy mugs. You need to work out that I own 12 mugs altogether before you do the sum.

Going up and going down

Working out increases and decreases with the Table of Joy is nice and easy. I know at least two ways to calculate increases and decreases using the Table of Joy, but I only give you one here.

To increase or decrease a number by so many per cent, here are the basic steps:

1. **Work out that many per cent of the original number using the Table of Joy.**

   For example, to increase 50 by 10%, work out 10% of 50 = 5.
   Alternatively, to decrease 200 by 5%, work out 5% of 200 = 10.

2. **To increase the number, add your answer from Step 1 to the original number.**

   In our example, add 5 + 50 = 55.

3. **To decrease the number, take your Step 1 answer away from the original number.**

   In our example, do 200 – 10 = 190.

Playing the Percentages

In this section, I look at the three most common uses of percentages and work through some examples with the Table of Joy. If you prefer using a different method to do your percentage sums, please use your way. As long as you get the same answer as I do, I'm happy.

Tax needn't be taxing

You use percentages a lot when you deal with tax. Politicians sometimes say 'pence in the pound' when they want to cover something up, but that means the same thing as per cent.

Your exam may contain a question like this:

> Clive earned £25,000 last year, before paying 20% income tax. How much did he send to the tax department?

This question asks you to work out 20% of £25,000. To answer this, you could set up a Table of Joy as in Figure 9-4, with your columns labelled 'number' and 'per cent', and your rows labelled 'part' (or, if you prefer to be clear, 'tax') and 'whole'.

You know the whole number is 25,000 and that corresponds to 100%, so you write those numbers in the appropriate squares on the bottom row of the table. You know the tax percentage is 20%, so you write 20 in the middle-right square. The Table-of-Joy sum is $25{,}000 \times 20 \div 100 = 5{,}000$. So Clive paid £5,000 in income tax.

Figure 9-4:
Clive's tax problem, solved with the Table of Joy.

	money	percent
tax		20
total	25000	100

$$\frac{25{,}000 \times 20}{100} = 5{,}000$$

A particularly sneaky examiner may ask you how much Clive had left after tax. To work this out, you take away the £5,000 he paid in tax from the original amount of £25,000, leaving Clive with £20,000. Make sure you read and understand the question.

A keen interest

You frequently see percentages when you look at savings and loans in your local bank. A typical exam question on this topic involves how much you have to pay on a loan, how much you receive in interest, or the interest rate on a loan or account. Here's an example:

> *Jerry borrows £5,000 to help him buy a car, and agrees to pay 5% per year in interest. How much does he owe at the end of the first year if he doesn't pay anything back?*

You need to work out 5% of £5,000 – the interest payment – and add that on at the very end. You can work out 5% of £5,000 with the Table of Joy as in Figure 9-5.

Whether you use the Table of Joy or another method, you find that 5% of £5,000 is £250. When you add that on to the original £5,000, you find that Jerry owes £5,250 at the end of the year.

Figure 9-5:
Jerry's car
loan.

	money	percent
interest		5
total	5000	100

$$\frac{5 \times 5,000}{100} = 250 \qquad \boxed{250 + 5,000 = 5,250}$$

Another question may work the other way around: the examiner gives you an investment and an interest payment (or dividend, or profit, or whatever), and asks you to figure out the percentage interest.

Here's an example of such a question:

> *Steve put £2,500 in a savings account. After a year, the bank paid him £75 in interest. What was the interest rate?*

Put everything into the Table of Joy as normal. The only slight difficulty is the sum: you have $75 \times 100 = 7,500$, which you then need to divide by 2,500. You can reasonably approach that by counting up 2,500s until you get to 7,500, which you reach on the third step, so $7,500 \div 2,500 = 3$.

Look back to Chapter 4 to see how to divide big numbers efficiently.

Changing prices

The last big arena for percentages is the shops. In the January sales, you can't move for signs saying '30% off everything' and 'Save 10%' on this, that or the other.

In the exam, you may need to work out the sale price of an item, or the new price after an increase, or even the percentage increase if you know the prices before and after. The first two of these sums are very similar. Here's an example:

> *My local stationer is offering 70% off calendars. The one I'm interested in, 'Beautiful Patterns of the Fractal World', was selling for £9.00 before Christmas. How much will it cost me now?*

I show you the Table-of-Joy version in Figure 9-6. To find the discount you do £9 × 70 = £630, and then divide that by 100 to get £6.30. Then you take that away from the original £9.00 to get £2.70 – which happens to be the exact amount of change I have in my pocket. Oh, happy days.

Figure 9-6:
Calendar discount with the Table of Joy.

	money	percent
discount		70
total	9	100

$$\frac{9 \times 70}{100} = 6.30$$

$$\boxed{9.00 - 6.30 = 2.70}$$

TIP

You could also do this sum directly by working out 30% of the original price, which is what remains after you take off 70%. That saves you a step of working but doesn't always give you a nice Table-of-Joy sum.

Working out a discount is also straightforward. Here's another example:

> *A nice coat is reduced from £80 to £56. What percentage is the reduction?*

Start by working out the price reduction: £80 – £56 = £24. Then set up a Table of Joy, as in Figure 9-7, and work out 24 × 100 = 2,400. 2,400 ÷ 80 = 30, so you have a 30% reduction.

Figure 9-7:
Coat discount with the Table of Joy.

	money	percent
discount	24	
total	80	100

$$\frac{24 \times 100}{80} = 30$$

Part III
Sizing Up Weights, Shapes and Measures

In this part . . .

Some mathematicians disagree with me, but I think the whole point of maths is to describe the real world – and that's what this part is about.

You probably know a lot of this already – I'd be a bit surprised if you'd never used a clock, a thermometer, scales or a ruler – but there are a few tricks and how-to's you need to have under your belt to be on top of the curriculum.

Chapter 10

Clocking Time

. . . .

. . . .

*Y*ou've probably used clocks since you were at primary school. In this chapter I help you get to grips with basic times and dates. I show you how to put theory into practice by reading timetables, doing sums with time, and even calculating speeds.

Understanding the Vocabulary of Time

Time is almost the only area of basic maths where there's never been any serious attempt to use a decimal system. (Angles, arguably, are another area.) So instead of using tens, hundreds and so on, time uses irregular units:

✔ 60 seconds make a minute

✔ 60 minutes make an hour

✔ 24 hours make a day

✔ 7 days make a week

✔ 365 days (about 52 weeks) make a year – except a leap year, which has 366 days.

Very short times – milliseconds – and very long times – millennia – do tend to work in tens. But at this level of maths, you're unlikely to need to know much about them.

A *century* is 100 years. The word 'cent' usually means something to do with 100 – there are 100 cents in a euro, and 100 centimetres in a metre. In cricket, a century is 100 runs.

A *millennium* (spelt with two Ls and two Ns) is 1,000 years. The prefix 'mill–' is often used in words connected with 1,000. I don't think anyone's ever scored 1,000 runs in cricket, but I imagine it would be called a millennium if they did.

Fractions in time

Time uses a few fractions. If you're not sure about fractions, check out Chapter 6, where I ease you through the subject.

The main fractions you meet in time are quarter- and half-hours. Half an hour is 30 minutes – because 30 is half of 60. Quarter of an hour is 15 minutes, because 15 is a quarter of 60. Three-quarters of an hour is 45 minutes, because 45 is three-quarters of 60.

You can write 30 minutes as ½ hour or 0.5 hours. Likewise, 15 minutes is ¼ hour or 0.25 hours, and 45 minutes is ¾ hour or 0.75 hours.

Be careful not to write '15 minutes = 0.15 hours'. That sum is not true, because there aren't 100 minutes in an hour. Fifteen minutes is a quarter of an hour and a quarter is 0.25. So you write: '15 minutes = 0.25 hours'.

Different date formats

To help with writing dates, each month has a number:

> 1 – January
>
> 2 – February
>
> 3 – March
>
> 4 – April
>
> 5 – May
>
> 6 – June
>
> 7 – July
>
> 8 – August
>
> 9 – September
>
> 10 – October
>
> 11 – November
>
> 12 – December

A very, very brief history of time

Our crazy, mixed-up system of time units is due mainly to the Sumerians (who lived about 4,000 years ago), who used 60 instead of 10 as their base for numbers. You can blame the same people for a circle having 360 degrees.

As far as I can tell, nobody knows for sure why a day is divided into 24 hours.

The European calendar, with its odd variations from month to month, is a variation on the Roman calendar (you may have noticed that some of our months are still named after Latin numbers).

A day is how long it takes the planet to spin around its axis. A year is how long it takes the planet to go around the sun – a little short of 365.25 days. The 0.25 is why we have a leap year every four years. The 'little short' is why things are actually more complicated than that.

Different countries use slightly different formats for writing the date. As an example, I use Stephen Hawking's birthday – 8 January 1942. In the UK, the convention is to write the number of the day, followed by the number of the month, followed by the year. So, Stephen Hawking's birthday is 8/1/1942 or – when there's no doubt about the century – 8/1/42.

In the USA, the convention is to write the day and the month the other way around – so 1/8/42.

Some science and computing fields write the date in yet other ways, but unless you work in one of these areas, you probably don't care.

Comparing the 12-hour and 24-hour clocks

When telling the time in the normal way, using the *12-hour clock*, we split the day into two halves – a.m., from midnight to noon through the morning, and p.m., from noon to midnight through the afternoon and evening.

The abbreviations a.m. and p.m. stand for the Latin phrases *ante meridiem* and *post meridiem*, which mean 'before midday' and 'after midday', respectively. If you say 'Let's talk at 9 o'clock tomorrow', the other person may not know which of the two 9 o'clocks you mean. Do you mean 9 a.m. (9 in the morning) or 9 p.m. (9 in the evening)?

To overcome this problem, travel timetables and the military often use the *24-hour clock*. Most digital watches, computers and phones have an option to switch between 12- and 24-hour clocks.

Instead of going back to one after midday, the hours keep on going up to 24 in the 24-hour clock. In the 24-hour clock, the time is always given by four numbers – so we write 1 p.m. as 1300 and we write 4.30 a.m. as 0430.

To say these out loud, you break the numbers into pairs and say 'hours' afterwards – 0430 is pronounced 'oh four thirty hours' and 1300 as 'thirteen hundred hours'.

Here's how to convert between the 12-hour and 24-hour clock:

1. **If your time is before 10 a.m., put a zero in front of the time. You're done.**

   For example, 6.57 a.m. becomes 0657, and 9.00 a.m. becomes 0900.

2. **If your time is between 10 a.m. and 12.59 p.m., don't do anything.**

   For example, 10.48 a.m. becomes 1048, and 11.00 a.m. becomes 1100.

3. **If your time is between 1 p.m. and midnight, add 12 to the hour.**

   For example, 1.02 p.m. becomes 1302, and 11.59 p.m. becomes 2359.

Going from the 24-hour to the 12-hour clock is easier. If the first pair of numbers is 13 or bigger, take away 12 to get the number of hours. Otherwise, you don't need to do anything.

So, 2300 is 11 p.m., and 1100 is 11 a.m.

Almost everyone gets confused by midday and midnight sometimes. The convention is that midnight is called 12 a.m. (either 0000 or 2400 in the 24-hour clock). Midday, on the other hand, is 12 p.m. or 1200. Personally, I like to make things clear by writing '12 midday' or '12 midnight' if there's any chance of confusion.

Catching the Bus: Seeing How Timetables Work

Timetables are a ruthlessly efficient way to get as much information as possible into as small a space as possible. This is why almost everyone gets confused by them, and many stations have those electronic signs telling you the 1715 to Weymouth is on time when it's already 1733 and you've seen no sign of a train for three hours.

A traditional bus or train timetable looks something like that in Figure 10-1 – compact and a little tricky to read unless you can focus.

					FX	FO
Appleborough	1100	1130	1200	1230	1300	1310
Bellstown	1112	1142	1212	1242	1312	1322
Colinsville	1119	1149	1219	1249	1319	1329
Doddsmouth	1139	1209	1239	1309	1339	1349
East Hill	1153	1223	1253	1323	1353	1403

Figure 10-1:
A traditional bus timetable.

Next time you need to understand a timetable, try following these steps (my example is for a bus timetable, but train timetables work just the same):

1. **Figure out in which direction you're travelling.** Some timetables list both directions – look for a timetable where your destination is listed below where you're travelling from.

2. **Find where you are and read across until you find a time that's after now (or around the time you want to travel).**

3. **Check any information at the top of the column to check the bus is running – you may see abbreviations such as 'FX' to say 'this bus doesn't run on Fridays' or 'WO' to say 'this bus only operates on Wednesdays'.** If that bus isn't running, find the next one.

4. **Put a finger on where you're travelling from and another on the time you've picked.** Move each finger down one stop at a time until you reach the stop where you want to get off.

5. **The time you're pointing at is when you ought to arrive at your destination.**

You use a similar method to figure out when you need to leave if you want to arrive somewhere at a particular time. Look at the arrival times for your destination and find an appropriate one. Put your finger on it and move up the column until you're at your departure point to find the time you need to leave.

Doing Sums with Time

Sums with time can catch anybody out. I got so fed up of adjusting my clock the wrong way and showing up two hours early or late for classes that for a while I stopped arranging to do anything on the days the clocks changed.

Any event has three main time-related properties: when it starts, how long it lasts, and when it finishes. If you know any two of these properties, you can work out the third – for example, if something starts at 11 a.m. and goes on for 30 minutes, then the event should finish at 11.30 a.m. But if time sums were always that easy, I wouldn't include a chapter about them in this book. Beware the nasty trap that comes up when you do sums with time that involve going into the next or previous hour.

For example, if it takes me 25 minutes to walk into town and I leave at 4.45 p.m., when do I arrive? If I add 25 to 45, I end up with 70 – but 4.70 p.m. isn't a real time!

I have at least three ways to avoid this trap. One way is to think slightly backwards: try to think of 25 minutes as '35 minutes less than an hour'. So, an hour later would be 5.45 p.m., and 35 minutes before that would be 5.10 p.m.

Another method is to split the 25 minutes into two parts – if I walk for 15 minutes, I get to 5 p.m., but I'm still 10 minutes away from town. So I add on another 10 minutes and arrive at 5.10 p.m. One other way is to accept that 4.70 p.m. doesn't exist as a time, but you know that 70 is 10 minutes more than an hour – so the time is four, plus an hour, plus 10 minutes, making 5.10 p.m.

Whenever you do a sum with time, be ultra-paranoid about going past the hour. Messing up time sums is surprisingly easy.

When does something start?

If you know when something ends and how long it takes, you can figure out when the event starts.

If the number of minutes in the duration is less than the number of minutes in the time, follow these steps to work out when the event starts:

1. **Convert the end time into the 24-hour clock.**

   For example, if you know your train arrives at 10.50 p.m., your end time is 2250.

2. **Take away the minutes in the duration from the minutes in the end time.**

   If your journey lasts for two and a half hours, take off 30 minutes, to get 2220.

3. **If you need to, take away the hours in the duration from the hours in the end time.**

   In our example, take off two hours, to get 2020.

4. What's left is the answer.

The train sets off at 8.20 p.m.

If the number of minutes in the duration is more than the number in the time, you need to follow this different approach:

1. Take the number of minutes in the duration away from 60.

For example, I know my drive to London takes an hour and 40 minutes and I want to arrive at 6 p.m. I take 40 away from 60 to get 20.

2. Add this number to the number of minutes in the end time. This is the number of minutes in the answer.

There are no minutes in the end time, so the number of minutes in my start time ought to be 20.

3. Take away the number of hours in the duration from the number of hours in the end time – and then take away one more. This is the number of hours in the answer.

I take away the 1 hour in my travel time from 18 (6 p.m. = 1800) to get 17 and then take off another to get 16, which is the number of hours in my departure time.

I need to set off at 1620, or 4.20 p.m. to get to London for 6 p.m.

When does something end?

If you know when something starts and how long it takes, you can figure out when it ends. Here's what you do:

1. Convert the start time into the 24-hour clock.

For example, the time is 12.45 p.m. and I want to cook a lasagne in the oven for 35 minutes. My start time is 1245.

2. Add the minutes in the duration to the minutes in the start time.

I add 35 minutes and get 1280.

3. Add the hours in the duration to the hours in the start time.

I don't have any hours to add.

4. If the number of minutes in your answer is 60 or more, add one to the number of hours and take 60 away from the minutes.

80 is 20 more than 60, so I get 1320.

5. What you have left is the answer.

My lunch is ready at 1320 – or 1.20 p.m.

How long does something take?

If you know when something starts and when it ends, you can figure out how long the event lasts.

If the number of minutes in the start time is less than the number of minutes in the end time, follow these steps:

1. **Convert the end time into the 24-hour clock.**

2. **Convert the start time into the 24-hour clock.**

3. **Take away the minutes in the start time from the minutes in the end time.**

4. **Take away the hours in the start time from the hours in the end time.**

5. **What's left is the duration.**

If the number of minutes in the start time is more than the number of minutes in the end time, you need to follow this slightly different approach:

1. **Take the number of minutes in the start time away from 60.**

2. **Add this number to the number of minutes in the end time. This is the number of minutes in the answer.**

3. **Take away the number of hours in the start time from the number of hours in the end time – and then take away one more. This is the number of hours in the answer.**

Speeding Along

Speed is a measure of how far you travel in a certain time.

Speed comes in a variety of units. Most UK cars measure speed in miles per hour (mph). In continental Europe, you generally see speed written as kilometres per hour (kmph or km/h). And scientists (just to be awkward) use metres per second (m/s). (A lecturer friend of mine insists that the imperial system using miles per hour is ridiculous and asks his students to work out speeds in furlongs per fortnight!)

The speed 60 miles per hour means that if you drive for an hour at that speed, you travel 60 miles. In half an hour you travel 30 miles. And in 2 hours you cover 120 miles.

To work out speed sums, I use the *Table of Joy*. Figure 10-2 shows the steps I use to work out what speed I have to travel at if I want to go 90 miles in three hours, following these steps:

1. **Draw a big noughts-and-crosses grid.**

2. **In the top-middle square, write 'Distance'. In the top-right square, write 'Time'.**

3. **In the middle-left square, write 'Speed'. In the bottom-left square, write 'Journey'.**

4. **In the middle-right square, write the number 1.**

5. **If you know the journey time, write that time in the bottom-right square (under 'Time' and next to 'Journey').** In this example, in Figure 10-2, you write 3. In other questions, you might have a time that's not a whole number of hours. If so, convert the time into decimals – so 1 hour and 30 minutes is 1.5 hours.

6. **If you know the distance of the journey (here, 90 miles), write the distance in the bottom-middle square (under 'Distance' and next to 'Journey').**

7. **If you know the speed, write the speed in the middle-middle square (next to 'Speed').**

8. **Write a question mark in the space that's left.**

9. **Write out the Table of Joy sum: times the question mark's neighbours together and divide by the other number. In this case, it's 90 × 1 ÷ 3.**

10. **Work out the answer. For this example, that's 30 miles per hour.**

Figure 10-2:
Using the Table of Joy to work out at what speed you need to travel to cover 90 miles in three hours.

	Distance	Time
Speed	?	1
Journey	90	3

$$\frac{90 \times 1}{3} = 30$$

Chapter 11

Working with Cold, Hard Cash

*M*oney is so deeply ingrained in our culture that defining money in a useful way is quite difficult. Basically, money is anything you can exchange for goods or services, and is used to value those goods or services. Money normally takes the form of coins and notes of certain, fixed values, but it can also exist only in the innards of a computer somewhere – when I pay my rent, a program in the depths of my bank changes something on a disk that makes me poorer and my landlord richer, although I never directly give my landlord either coins or notes. Money also takes different forms (currencies) in different countries. You can get into all sorts of philosophical arguments about whether money actually *is* the notes and coins, or simply what they represent, but here is not the place for that discussion.

In this chapter, I give you a brief introduction to money as used in the UK and the sums you need to be able to do in the normal course of life. For this reason, this chapter is useful both for day-to-day living and for any maths exams you hope to take.

I show you how to use the Table of Joy to convert money from one currency to another, so you can splash your cash on holiday. I also run you through some of the more complicated money sums you may need to tackle at some point.

Seeing What You Already Know

The Sumerians invented money about 4,000 years ago as a way to make trading easier. By using money, if you owned a cow and wanted a haircut, you didn't need to find a barber who wanted a cow – you could simply sell the cow to someone and then use the money for a short back and sides.

Money sums work like normal sums. Most money sums have a decimal point, but you still add, take away, times and divide with money, just like you do with regular numbers.

To show that a number represents money, in most countries you put a currency marker before the number. For example, to show ten pounds, you write £10. If you're dealing with dollars instead, you write $10.

The £ is a fancy letter L, which comes from the old Latin name for pounds, *libra*. The euro symbol, €, is just a fancy E. The dollar sign, $, isn't a fancy S but is probably based on the shape of Spanish doubloons – the 'pieces of eight' that pirates and their parrots jabber on about.

Counting coins and notes

The UK has a pretty limited range of coins: the small-value copper coins (the 1p and the bigger 2p); the medium-value silver coins (the 5p and the bigger 10p are round, while the 20p and the bigger 50p are seven-sided); and the higher-value gold coins (the £1 and the bigger £2, which has a silver bit in the middle).

The UK uses the following notes: the £5 blue note, the brown £10 note, the purple £20 note and the red £50 note. (The Eurozone also has multicoloured money. In the US though, all the notes are shades of green.)

Exploring examples of money sums

Here are some typical things you may need to work out using money:

- ✔ **Calculating commission:** Working out how much a sales rep gets paid.

- ✔ **Changing money:** Translating prices from one currency to another.

- ✔ **Deciding what you can afford:** How many things can you buy with the money you have available?

- ✔ **Shopping in the sales:** Adjusting prices to take into account special offers.

- ✔ **Totalling how much a group of things costs:** Say, a basket of shopping or a series of journeys.

- ✔ **Using rates and fixed charges:** For example, hiring equipment or booking a hotel.

- ✔ **Working out a deposit:** Using a percentage to figure out how much you need to spend up front.

You can probably think of lots of other examples. The list above gives you an idea of the questions you may see in a maths exam.

Running the Numbers

To solve any maths problem correctly, it's a good idea to pause for a moment and think about the question. It sounds obvious, but I see a lot of students throw away a lot of marks by trying to do a sum before they really know what the sum is. Particularly with money questions, it's tempting to think 'this is easy' and dive in – but that's a sure-fire way to get into trouble.

To avoid this terrible outcome, you need to follow these three steps:

1. **Work out what the question is asking.**

2. **Gather the information you need.**

3. **Do the sum.**

A lot of students skip the first two steps above and wonder why they get the third step wrong. Don't be like them.

Particularly with money questions, take a few moments before you start scribbling down sums and make sure you understand the question. Do you need to find the difference between two amounts? Are you trying to find a multiple of a price?

Apply some common sense to this kind of sum and ask: 'Should my answer be more or less than the value I already know?'

Adding and subtracting money

The simplest sum you can do with money is to add up how much a bundle of goods costs – for example, all the stuff you buy in one shop, your total bills for the month, the cost for a meal in a restaurant, or the cost of a trip somewhere. Adding with money is exactly the same as adding decimals, which I cover in detail in Chapter 7. Here's what you do:

1. **Make sure all of your numbers are in pounds and pence.**

   If you have a number in whole pounds, write '.00' after it.

2. **Write down all of the numbers in a big list, lining up the dots.**

   Writing your columns neatly makes adding up the numbers easier.

3. **Add up the numbers in the normal way.**

   You don't need to do anything special with the dots.

4. **Put a dot in your answer, in line with all of the other dots.**

Your exam's unlikely to have a question as simple as 'Add up these numbers'. You may first need to work out the amounts you need to add up, perhaps by multiplying a daily hotel rate by how many days a guest stays.

Taking away with money is just the same as taking away with decimals. Follow these steps to take away one amount of money from another:

1. **Make sure both of the numbers are in pounds and pence.**

2. **Write the numbers down, with the bigger number at the top, and the dots lined up.**

3. **Take the numbers away as normal, ignoring the dot.**

4. **Put a dot in your answer, in line with the other dots.**

To take away several numbers from one number, you have two options: you can do a series of take-away sums, or you can add up all of the costs and then do a single take-away sum.

Multiplying and dividing money

If you go into the 99p shop and buy five items, you have to pay 99p five times over, or 5 × 99p. When you buy several identical things, you do a times sum to get the total amount you need to pay.

Multiplying with money is just like working with normal decimals (I cover the main 'points' of decimals in Chapter 7). Here's how to do a times sum with money:

1. **Make sure your money number is in pounds and pence, and write it down.**

2. **Times the number by however many you need to times it by, using your favourite method from Chapter 4.**

   Ignore the dot at this point.

3. **Put the dot in place, directly below the dot in the original money number.**

You use divide sums in money when you want to split an amount of money between two or more people, and when you know the price of several items but need to know the price of just one. Dividing with money works just the same way as for normal numbers, as I show in this recipe:

1. **Write down the number you want to divide, making sure it's in pounds and pence.**

   Give yourself plenty of space between the digits.

2. **Draw a 'bus-stop' over the number – a vertical line before the start of the number and a horizontal line over the top.**

3. **Write the number you want to divide by in front of the bus-stop.**

4. **Divide as normal, ignoring the decimal point.**

5. **Put the dot in place, directly above the dot in the original number.**

Rounding money numbers

Sometimes you need to estimate money sums. As with everything else in this chapter, rounding sums with money works just like rounding sums with any other numbers – you round off to the nearest numbers that are easy to work with, and then do the sum as normal. Check out Chapter 5 for the low-down on rounding.

Here are some day-to-day examples of when you may want to round money sums:

✔ Rounding the prices of a number of individual goods to the nearest pound or ten pounds before adding them all up.

✔ Checking an answer using estimated values, for example, to work out whether a calculated commission is roughly correct. I cover this in detail in 'Choosing the right sum', later in this chapter.

✔ Finding an approximate, simple ratio of two amounts – say, the ratio of £2.99 to £11.99 is roughly the same as 3:12 or 1:4.

✔ Estimating percentages. An exam question might ask 'Roughly what percentage is £3.80 of £39?' To solve this, you round the £3.80 to £4 and the £39 to £40 and work out that it's roughly 10 per cent.

Using Euros, Dollars and Other Currencies

Different countries use different currencies. The UK uses the pound (or 'pound sterling'). The USA uses the dollar, not to be confused with the Australian dollar (used in Australia, unsurprisingly) or any of the other dollars used worldwide. Most of Europe uses the euro, although a few countries still use their own currency out of choice or because they haven't qualified to join the euro.

Exploring exchange rates

Different currencies have different conversion rates between them. These rates change from day to day – but at any given time, the rates all magically match up so you can't make money by endlessly changing your funds from one currency to another.

In principle, I can take a £20 note, go to the bank, and ask the bank to change the pounds into euros. The bank gives me about €23.70 in exchange for my £20. (The exact rate will have varied by the time you read this.) I then change my mind and decide to go to the US instead, and ask the bank to convert the euros into dollars. The bank gives me about $32. Then I decide to stay at home and change all my money back into pounds and get my £20 back – before the bank kicks me out for being a nuisance.

 Currency exchange doesn't work quite like I describe above. Every time the bank, travel agent or bureau de change changes your money, it uses an exchange rate that's slightly better for them and slightly worse for you – so it makes a small profit on each transaction. Therefore, if I went to the bank and pulled exactly the stunt I mention, I'd probably leave with £5 less than I started with.

Converting currency with the Table of Joy

When you travel to another country, you need to convert some money into the currency used in that country. Normally, you go to the bank, travel agent or post office and say 'I'd like to convert this money into euros, please' (or whichever currency you need). The cashiers work out the amount for you, but it's worth checking that they haven't made a mistake.

The Table of Joy helps you do currency conversion without tears. Head to Chapter 8 if you haven't met the Table of Joy yet. Here are the steps you take to make sure you're not at the bureau de short-change (I give some examples in Figures 11-1 and 11-2):

1. **Draw a noughts-and-crosses grid.**

   Make the grid big enough for you to write labels in.

2. **Label the columns with the names of your two currencies.**

   For example, write 'pounds' and 'dollars'.

3. **Label the rows 'exchange rate' and 'money'.**

   Or invent your own labels to help you remember which row is which.

4. **Write the exchange rate in the 'exchange rate' row.**

   If the exchange rate is £1 = $1.60, write 1 in the pound column and 1.60 in the dollar column.

5. **Write the amount of money you know about in the appropriate column of the money row.**

   If you want to change £200, write 200 in the pounds row. Put a question mark in the remaining square.

6. **Do the Table of Joy sum: times together the number above and the number beside the question mark, and then divide by the remaining number.**

   Your answer is the converted amount of money. In this case, it is be $200 \times 1.60 \div 1 = 320$.

The exchange rate between pounds and dollars is £1 = $1.60
Alice wants to change £200 into dollars.
How many dollars does she receive?

	Pounds	Dollars
Rate	1	1.6
Money	200	?

$$\frac{1.6 \times 200}{1} = 320$$

Figure 11-1: Converting pounds into dollars.

The exchange rate between pounds and euros is £1 = €1.20
Bernie wants to change €360 into pounds.
How many pounds does he receive?

	Pounds	Euros
Rate	1	1.2
Money	?	360

$$\frac{1 \times 360}{1.2} = 300$$

Figure 11-2: Converting euros into pounds.

Managing More Complicated Money Sums

Many people are more comfortable working things out with money than they are doing sums with plain numbers. For this reason, maths test questions involving money are often a bit harder than other sums.

For example, you may need to work out the cost of each instalment on a piece of furniture, or the initial deposit you have to pay to secure the furniture. This kind of sum is fiddly rather than difficult – and thinking about the sum in terms of money rather than random numbers makes a lot more sense.

Another favourite of maths examiners is the amount of commission a sales rep earns over a total number of sales. This is really just a percentages question in disguise, but in this section I run you through the commission sum, in case you haven't read Chapter 9 on percentages yet.

Also in this section I consider how to choose the right sum. Questions about the 'right sum' are both easier and harder than other questions in your maths test – easier because you don't have to work out the answer, but harder because your way and the examiner's way of doing a sum may not be the same, so picking the 'right sum' from a list of options isn't easy.

Dealing with deposits and payment plans

When you buy something expensive – a car or a piece of furniture, for example – you may have the option to pay in instalments, perhaps after putting down a deposit. In this section I help you work out how much deposit to pay, and how much each instalment costs.

Deposits

You normally describe a deposit as a percentage or fraction of the total cost. The first thing to do is work out the total price of the item you want to buy. In a maths test, the examiner may tell you the total price; if not, you have to work it out using a table or some kind of sum. I cover working with tables in depth in Chapter 16.

You may need to work out the total cost from a table that tells you about fixed charges and daily rates – for example, if you hire a van, you might have to pay an initial fee and then a charge for each day.

To work out the total cost, you multiply how many days you need at each rate by that rate, then add up all of the answers. Add on any fixed charges you need.

After you figure out the total cost of the item, you can work out the deposit using a percentage sum. I cover percentage sums in detail in Chapter 9, and give an example in Figure 11-3, but here's a quick guide of what to do:

1. **Draw a noughts-and-crosses grid.**

   Leave plenty of room for labels.

2. **Label the columns 'money' and 'per cent', and the rows 'deposit' and 'total'.**

3. **Write the total cost in the 'money/total' cell.**

   That corresponds to 100 per cent, so write '100' in the 'per cent/total' cell.

4. **Write the deposit percentage in the 'per cent/deposit' cell.**

   Put a question mark in the 'money/deposit' cell.

5. **Write down the Table of Joy sum (check out Chapter 8 for more on the Table of Joy).**

   Times the neighbours of the question mark together ('money/total' × 'per cent/deposit' ÷ 'per cent/total').

6. **Do the sum.**

Steve buys a television that costs £700.
The shop requires a deposit of 15%.
How much does Steve have to pay as a deposit?

Figure 11-3:
Working out
a deposit
from a
percentage.

	Money	Percent
Deposit	?	15
Total	700	100

$$\frac{15 \times 700}{100} = 105$$

A question may also tell you that a deposit is a certain fraction of the full price. Fraction sums aren't really that much different from the percentage sums you just did: you label the right-hand column of the Table of Joy as 'fraction'. Put the bottom number of the fraction in the 'fraction/total' cell and the top number of the fraction in 'fraction/deposit' cell. Then the calculation is exactly the same.

A percentage is a fraction with 100 on the bottom. So don't be surprised that the Table of Joy works for both percentage and fraction sums.

Payment plans

Figuring out the size of a regular payment is one more step in the process I describe in the previous section. Follow these steps to work out your payment plan:

1. **If you know you have to pay a deposit, work out the cost of the deposit as I describe above.**

2. **Take away the cost of the deposit from the total cost.**

3. **Divide what's left over by the number of payment periods.**

   The answer is the size of each of your regular payments.

Calculating commission

When you go into an electrical goods shop and find yourself swarmed by salespeople keen to be your best friend, they may be making a misguided attempt to give you good customer service, but more often than not they're working on commission and want you to buy something. Salespeople on commission get a bonus corresponding to a percentage of what you spend in the shop.

Commission isn't restricted to electronics salespeople, of course – sales reps in general, employment agencies, authors, sports stars' agents . . . all of them get a cut when someone buys their product.

In a commission question in a maths test, the question usually tells you the total amount and the percentage commission the agent receives. The sum you need to do is a regular percentage sum – the steps are exactly the same as in the previous 'Deposits' section.

Some commission questions are a little more complicated. For example, you may need to work out how much a sales rep earns who is paid a base salary (per hour, per day or per week) along with a bonus based on their sales figures. Here's how you deal with such a situation:

1. **Figure out how many hours (or days or weeks) the rep works.**

2. **Times this by the relevant payment rate to get the rep's base salary.**

3. **Look at the commission structure and decide what number you need to know.**

   Commission is often based on the total number of sales or the total amount of money.

4. **Read the commission structure carefully and decide which sum you need to do to work out the commission.**

   Usually you do a percentage sum (like the one in the 'Deposits' section earlier) or a simple divide.

5. **Do the sum.**

6. **Add the result from Step 5 to the base salary from Step 2.**

   That's the total amount the rep earns.

Sussing out sales prices

Working out the price of something in the sales is one of the most common areas people use maths in real life. Sales prices are normally given as a percentage or a fraction off: for example, a sign may say 'reduced by 20%' or '⅕ off'. To work out the price, you do the same sum as in the deposit and commission sums above, and then add an extra step:

1. **Work out either the fraction or percentage of the original price as you did in the 'Deposits' section.**

2. **Take this number away from the full price.**

That hardly seems worth a recipe, does it?

Choosing the right sum

'Which of these sums gives the correct answer?' is a favourite question of maths examiners. Unfortunately, the question isn't a great favourite of mine or of anyone I know.

As many 'correct' sums exist as there are types of question – in fact, many more, because my way of doing a sum correctly may be different from your correct way, and both of these may be different from the examiner's way.

Here are some typical sums you may be asked to deal with when picking the 'right' money sum:

✔ Calculating or estimating a commission.

✔ Checking takings, totals and expenses are correct.

✔ Working out an original price if you know the discount.

✔ Working out a price after a discount or sale.

Too many possibilities exist for me to show you how to do them all. Instead, in this section I give you some general strategies for working through this kind of question in a maths test.

If you struggle with this kind of question in the exam, miss it out first time through and come back to it at the end if you have time. This is generally a good idea for any question you find tricky.

I can think of three good strategies for attacking the 'Which calculation?' type of question: think about how you'd solve the sum, translate the answers the examiner gives, and try a few numbers to see what happens.

Thinking about how to solve the question

The most 'mathsy' way of answering a 'Which sum?' question is to consider how you'd do the sum yourself. For example, if you have a percentage question, see what the Table of Joy gives as the sum. If you have a sales price, think about the sum you need to do – will the answer be more or less than the price you already know? How will you work out the discount?

This method largely involves using the techniques I describe throughout this chapter and writing down the sums you need to do.

The order in which you times or add things doesn't matter – so if your sum is 15×100 but the question offers 100×15 as an answer, consider the two sums the same. Your answer isn't necessarily wrong just because it doesn't look identical to the answer given.

Translating the answers provided

If you're completely stumped about how you'd go about solving or checking an answer, try to decipher what each of the given answers means. This can be a little bit time-consuming, but it's good revision practice – getting familiar with intimidating sums can help you feel less afraid of them when the exam rolls around.

For example, you may look at a sum containing '÷ 100' and say, 'I know that means the same thing as "per cent" – does that make sense in this context?' You may also ask, 'Does this make the number bigger or smaller, and which way should it be going?'

Throwing in numbers

The most brutal of the methods I describe in this section is simply to work out the number you get doing the sum your way and then compare your number with the number in each of the optional answers.

This is a lot of work, but it's also a really good way to get a grip on how the sums work.

If you can dismiss an answer or two as 'impossible' before you start working things out, you can cut down the amount of work you need to do in 'Which sum?' questions.

Chapter 12

Taking the Weight Off Your Shoulders

You probably know quite a bit about weight already. For example, you know that moving a heavy object is harder than moving a light object. You may know one or two people who obsess over how much they weigh – sometimes to an unhealthy degree. And you probably notice that you buy most food by weight and follow recipes according to weight – for example, '200 grams of the finest shiitake mushrooms known to humankind' or '1 kilogram of sugar' if you're following my mum's fudge recipe.

In this chapter I look at some of the obvious and not-so-obvious facets of the maths of weight. I show you how to use different types of weighing scale. I introduce the different units you use to weigh things, explain how to convert between different units, and offer some guidelines on how to estimate.

I also tell you about the one thing maths examiners like to question more than anything else – the maths of cookery – so I take you through scaling recipes and working out the money side of weight and cooking times.

Appreciating What You Already Know

At some point in your life, you have probably stepped on some bathroom scales. If you're anything like me, you stepped straight off again in horror and loudly proclaimed the scales to be broken.

Pretty much anything you use to measure weight is called a scale (or a weighing scale). You can still see the earliest type of scale (a *balance*) in antique shops and in statues of Justice: a balance comprises two trays that balance on or hang from a pivoted beam. You place what you want to weigh on one tray and place a known weight on the other tray. When the beam is level, you have the same amount of weight in each tray. Antique scales normally come with various lumps of metal marked with the appropriate weight. Surprisingly, a well-designed balance is the most accurate tool we have for measuring the mass of an object.

A step up from the balance in ease of reading is the *analogue scale*, which is more like the scale in my bathroom or the scales you see in the veg section of the supermarket. These scales have only one surface, where you put the thing you want to weigh. A dial rotates to show the correct weight. By far the easiest scales to use are *digital scales* – some bathroom scales look like this, as do the things you put your suitcase on at the airport before the check-in person says you have 1kg too much luggage and charges you an extra £50. More pleasantly, the scales in my local coffee shop are digital. Digital scales have one surface to put things on. The weight appears as a number on the screen. I have no idea how they work. If you find out, tell me! I show several types of scale in Figure 12-1.

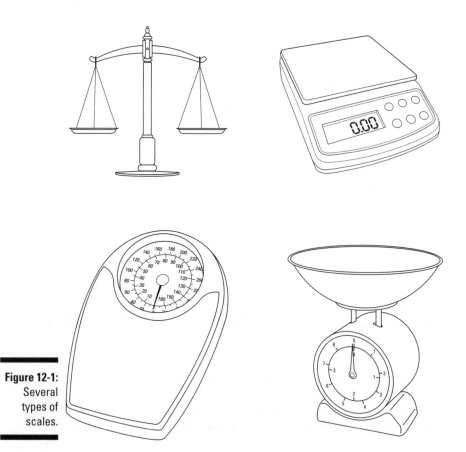

Figure 12-1:
Several
types of
scales.

Using digital scales

I used to own a cookbook called *How to Boil an Egg*. It's the kind of thing you probably look at and think 'How could anyone *possibly* get through adolescence without knowing that?' As it happened, I did know how to boil an egg, but I wasn't very good at it – and I'm still not. But I'm conscious that things that seem trivially easy to some people are quite hard for others. So, if using a digital scale is obvious to you, fine – skip to another section. I still want to help the people who are a bit puzzled by it.

A digital scale is a scale with an electronic display, a bit like a calculator. To weigh something, here's what you do:

1. **Check that the display reads zero before you start.**

   Those smarty-pants who skipped ahead may not realise you need to do this – so thanks for sticking around. If the display doesn't read zero, you might need to 'zero' the scale by pressing a button.

2. **Put what you're weighing on to the weighing surface – the plate or tray or bowl.**

3. **Read the number off the display.**

Don't forget to note the unit as well as the number. Some scales measure in kilograms and grams, but others measure in pounds and ounces. Make sure you notice the decimal point if there is one.

Using analogue scales

Analogue scales were common when I was growing up (I still think a bathroom's posh if it has a digital scale in it). Analogue scales contain an ingenious system of springs inside that measures how much weight you apply to the weighing surface, which rotates a dial marked with the appropriate weights until the number underneath the hairline on the display shows the weight you want to weigh.

As far as you need to know, analogue scales work by magic – I've always considered this to be the case and it's done me no harm at all. All you care about, unless you have a job as a scales technician (in which case, I expect you already know all there is to know about weighing things), is how to adjust and read an analogue scale. So, let me tell you:

1. **Make sure the scale is set to zero.**

   Without putting anything on the scale, look at the hairline and make sure it's over the zero mark on the dial. If not, twiddle the little twiddly thing at the front of the scale until the zero matches up with the line.

2. **Put the thing you want to weigh (possibly yourself) on the scale.**

3. **Look at the display and see what's underneath the hairline.**

   If the hairline lies over a tick marked with a number, that number is your measurement. If you're not on a tick, you need to estimate the weight.

   Count how many ticks there are between ticks with numbers by them, figure out how much weight each tick represents, and count how many ticks above (or below) a known point you are. You then times the number of ticks by the size of each tick, and add that to (or take it away from) the value given by the numbered tick.

Most scales have at least two sets of numbers on the dials – one for imperial measurements such as pounds and ounces, and one for metric measures such as kilograms and grams. Make sure you look at the right one.

Figure 12-2 shows some examples of reading a dial. The left-hand scale shows about 73kg and the right-hand scale about 440g.

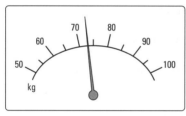 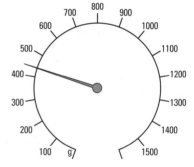

Figure 12-2:
Reading
a dial.

Using a balance

An old-fashioned balance is the most difficult of the three types of scale to use – but also the most satisfying . . . maybe just because I'm a sucker for doing things the hard way.

Using a balance is very unlikely to come up in a maths exam, but you may need to weigh something in your mother-in-law's kitchen one day . . . and you don't want to show yourself up in front of your mother-in-law.

Here's how to use a balance:

1. **Make sure the pans are level with each other before you start.**

   If they're a little bit off, try sorting them out with a well-judged tweak. If they're very off, I can't help. Maybe ask your mother-in-law why her scales are off-balance.

2. **Put the thing you want to weigh in one of the pans.**

3. **Add the biggest weight you haven't tried yet to the other pan.**

4. **If the weight side drops, take that weight off and put it to one side.**

5. **Go back to Step 3 until the pans balance.**

6. **Add up the weights in the pan.**

The answer is the total of the weights in the other pan.

Measuring Weight

Reading measurements off of a scale is easy enough – but the different systems used on different scales can be confusing. In my mum's kitchen, there's a lovely antique set of scales with weights marked in ounces, which was more or less useless any time I wanted to cook something at home because all of my recipe books called for a certain number of grams.

The choice of whether to use imperial units or metric units can be a controversial one – I can think of some famous cases of market traders being fined for refusing to sell their produce in kilos – but I have a general attitude that whatever system makes the sums easier is the best one for me to use – in this case, I can't think of an example where that wouldn't be the metric system.

Your maths exam and pretty much anything you do at work are almost certain to use metric (possibly with an imperial equivalent) – so I use metric.

The metric system: Grams, kilograms and tonnes

The base unit of mass is the kilogram – written 'kg' and sometimes called 'a kilo'. A kilogram is almost exactly the same mass as $1,000 \text{cm}^3$ of water. If you weigh a one-litre bottle of water, it should be just a smidge more than 1 kilogram, depending on how heavy the bottle itself is.

'Kilo' means thousand, so a kilogram is made up of 1,000 grams. A gram is a pretty light weight, maybe about the same as one of the small packets of sugar you might find on the table in a cafe.

Going the other way, a tonne is 1,000 kilograms, about the same weight as a small car, or (almost exactly) a cubic metre of water.

The imperial system: Ounces, pounds and stones

You only really need to know that the imperial system exists, rather than understand its precise details. For the sake of completeness – and in an attempt to convince you of the inherent superiority of the metric system – I run through some of the imperial basics in this section.

The imperial system has a certain illogical charm to it: a stone is made up of 14 pounds, and a pound is made up of 16 ounces. That means a stone is 224 ounces. (Do you see why I prefer to work with the nice tens, hundreds and thousands of the metric system?)

An ounce works out to be about 28 grams – which you normally round to 25 grams or 30 grams when cooking. An ounce is about the same weight as a big handful of rice. A pound is a little less than half a kilogram, about the same weight as a regular tin of tomatoes. A stone is a bit more than 6 kilograms, about the same as a bowling ball.

Converting Weights

In real life, you may need to convert between different weight systems – for example, American recipes (and old-fashioned British ones) generally use pounds or ounces for weights, and your scales may only show kilograms or grams.

You have three main ways to convert weights: using a conversion table, using a graph and using a formula – in roughly that order of difficulty.

Using a table

Converting weights using a table is as easy as looking up numbers in a chart. You find the column showing the unit you have and look down that column until you find the weight you have. Then you find the other number in the same row, and write it down.

If your number isn't in the table, the first thing to do is check you're looking in the right column. (I mess that up all the time.)

If you're sure the number isn't there, you need to be a little smarter. Here are some things you can try:

✔ **Is your number about half-way between two other numbers, or some other fraction you can spot?**

Give an answer somewhere between the two corresponding numbers that makes sense.

✔ **Can you make up the weight you're looking for by adding other weights together?**

Try adding together the corresponding weights in the other column.

✔ **Is there any other information in the table that you can use?**

For example, is it double, triple or ten times as big as another weight that's listed?

✔ **Are you totally sure the number isn't in the table?**

Close your eyes, take a breath and look again.

Using a graph or a chart

You can use a graph or chart to convert weights. The two methods are similar, with some slight differences.

A graph in this context is almost certainly a straight-line graph (flick to Chapter 16 to find out more about reading graphs). Along the bottom (horizontal) axis you see numbers corresponding to one weight unit. On the left-hand (vertical) axis you see numbers corresponding to the other weight unit. The graph has a line showing how the two are linked together.

To use a graph to convert weight, here's what you do:

1. **Work out whether the number you have belongs on the horizontal or vertical axis by looking at the units.**

   Whichever axis has the same units as you have is the correct axis.

2. **Find the number of your weight on the axis.**

   If you can't see your number, you may need to do a little estimating: find the two marks your number comes between and decide whether your number's midway between them or closer to one number or the other. Make a mark where you reckon your number should be.

3. **If you're on the horizontal axis, draw a line (use a pencil and ruler, and draw lightly) vertically upwards until you reach the graph line.**

   If you're on the vertical axis, draw a line horizontally to the right until you reach the graph line.

4. Now change direction and draw a line to the other axis.

If previously you drew upwards, use a pencil and ruler to draw left from the point on the graph line you reached until you reach the vertical axis.

If previously you drew to the right, use a pencil and ruler to draw a line downwards until you reach the horizontal axis.

5. Read off the value you reach.

If the value isn't bang on a mark, do some intelligent guesswork to get the value. Think about what numbers the value sits between and judge whether it's closer to one than the other. The number you come up with is your answer.

I show you how to convert weights using a graph in Figure 12-3.

Figure 12-3: Reading a weight graph. To convert 80kg into pounds, read up from 80 to the line and across, to get around 180lbs. To convert 300lbs into kilograms, read across from 300lbs to the line and down, to get about 135kg.

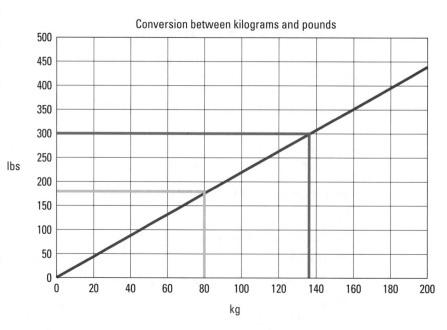

Conversion between kilograms and pounds

Instead of a graph, you may have a chart – the chart may look a bit like a thermometer, with two scales and numbers marked on each, as in Figure 12-4. Follow these steps if you want to convert weights using a chart:

1. **Decide which side of the chart is the correct one for the measurement you have.**

   Look at the unit (kilograms? ounces?) and find the right unit on the chart – the unit is often at one end or the other of the chart.

2. **Find the number you have on the same side of the chart as the unit you have.**

   If the number isn't there, make an inspired guess about where it should be: find the numbers your number fits between and judge whether it's halfway between them or closer to one than the other. Make a mark on the chart where you reckon the number should be.

3. **Look at the other side of the chart and read off the number that lines up with your mark.**

   You may need to do a bit of guesswork if you're not exactly on a numbered mark. Are you midway between two marks or closer to one than the other? Come up with a number that feels about right.

In a multiple-choice exam, you can always look at the possible answers and choose which one looks like it's closest to your mark. Eliminating the impossible is a perfectly good way to do maths!

Figure 12-4:
Using a chart to convert weights.

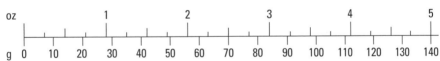

Using a conversion factor

You can use a *conversion factor* to change a weight from one unit to another. A conversion factor is simply an equation linking two kinds of weight – for example, '1 ounce equals 28.5 grams' or '1 kilogram equals 2.2 pounds'.

The numbers you use to convert between metric and imperial weights are very rarely nice, round numbers, but the actual process of conversion is pretty straightforward.

The Table of Joy, which I explain in detail in Chapter 8, takes a lot of the thought process out of deciding which sum you need to do to convert weights. Here's how to use the Table of Joy to convert between units of weight:

1. Draw a noughts-and-crosses grid.

Leave plenty of space for labels.

2. Label the columns with the two units.

This may be ounces and grams, or kilograms and pounds, or various other combinations.

3. Label the rows 'conversion factor' and 'weight'.

4. Fill in the cells as appropriate.

Write the conversion factors in the conversion-factor row. If the question tells you that 1 kilogram = 2.2 pounds, write '1' in the kilogram column and '2.2' in the pounds column. Put the weight you have in the correct column and put a question mark in the remaining cell.

5. Write out the Table of Joy sum.

Times the number in the same row as the question mark by the number in the same column as the question mark, and divide by the number opposite.

6. Work out the sum and write down your answer.

One of the strengths of the Table of Joy is that the table works either way around, regardless of which unit you already know. Whether you know the weight in ounces or grams, or in pounds or kilograms, if you fill out the table correctly you can still determine the right sum to do.

If you feel brave, you may want to try the following other ways of converting weights:

✔ Look at the unit of the weight you have. If that unit has a '1' by it in the conversion factor, you *multiply* your weight by the other number.

✔ If the unit of the weight you have doesn't have a '1' by it, you *divide* your weight by that number.

For example, your maths test may contain the following question:

You are told that 1 pound is about 450 grams. Seven pounds is the same as half a stone. How many grams is half a stone?

You can work out the answer like this:

1. If 1 pound is 450 grams, 7 pounds is seven times as much.

2. $7 \times 450 = 3{,}150$.

3. Half a stone is about 3,150 grams (or 3.15 kilograms).

Estimating weight

Having an idea of how much things weigh can be a real advantage when you want to do weight sums. For example, if you work out that an elephant weighs 15 grams, you'll want to know that's preposterously light and your sum's gone wrong. Here's a table of some common things and roughly what they weigh, so you can start to get an idea of whether your numbers are on the right lines:

5p piece	About 5g
Handful of rice	About 25g

Apple	About 100g
One-litre bottle of water	1kg
Big bag of potatoes	10kg
Typical adult male	80kg
Adult bull	500kg (or more)
Medium-sized car	1,000kg (a tonne)

Always give your answers a 'sanity check' before you write them down. Estimating is a good tool for this, as is having an idea of how your answers should fit into the real world.

Alternatively, use the Table of Joy, as I show you in Figure 12-5. Using the table is slightly overkill for this example, but until you've used the table a few times you won't know it's overkill.

Figure 12-5:
Weight conversion using the Table of Joy.

	Pounds	Grams
Conversion	1	450
Weight	7	?

$$\frac{450 \times 7}{1} = 3150$$

Weighing in Cookery

One of the places you're most likely to weigh things is in the kitchen. If you look at almost any recipe, it calls for a few ounces of this or a certain number of grams of that.

The numeracy curriculum expects you to know the following food-related skills:

✔ **Buying by weight:** Your ingredients need to come from somewhere. I show you how to work out the price and value of things when you buy by weight in the next section.

✔ **Comparing value:** Still in the shops, if you have two different sizes of box, which is better value?

✔ **Cooking by weight:** The bigger – read heavier, in this context – something is, the longer it takes to cook. Your maths test may contain a question that gives you an equation or guideline for cooking by weight and ask you to figure out the cooking time.

✔ **Recipe scaling:** If you have a recipe that serves four people, what do you have to do to feed 16 people? This kind of question comes up all the time in maths tests. You can solve these questions using the Table of Joy (check out Chapter 8 for more details on this).

✔ **Weighing:** You should be able to read digital and analogue scales to determine how much of an ingredient you have. Look at the sections 'Using analogue scales' and 'Using digital scales' earlier in this chapter to see how to weigh stuff in your kitchen.

Buying by weight

When you buy something by weight – say, fruit and vegetables – your receipt gives three pieces of information next to the name of the item:

✔ The *weight* you bought, in kilograms or grams.

✔ The *price per kilogram* of the item.

✔ The *total price* of what you bought.

If you have any two of these bits of information, you can figure out the other one. You can do these sums easily using the Table of Joy.

In your maths test you'll probably only need to work out the total price, based on the other two bits of information. But being able to do all three sums is handy. If you know the weight and the price per kilo, you work out the total price by multiplying the two together. To work out the price per kilo or the weight you've bought, I suggest you follow these steps using the Table of Joy:

1. **Draw a noughts-and-crosses grid.**

   Leave plenty of space for labels.

2. **Label the columns 'money' and 'weight'.**

3. **Label the rows 'price' and 'total'.**

4. **Fill in the numbers you can.**

   The weight of your items goes in the 'weight/total' cell. The 'weight/price' cell is 1. The 'money/price' cell is how many pounds stirling per kilogram you have. The 'money/total' cell is how much the total comes to. One of these will be missing – fill that cell with a question mark.

5. **Write the Table of Joy sum.**

   Times the number in the same row as the question mark by the number in the same column as the question mark, and divide by the remaining number.

6. **Work out the sum.**

   This is your answer.

If you have all four of the numbers in Step 4, the sum is already done.

Comparing value

You can frequently see me in the cereals aisle of the supermarket holding two different types of muesli trying to work out which is the better *value*. As I stand there in your way, I work out which muesli has a cheaper price per kilo.

I could, of course, just look at the sticker on the shelf that gives the price per kilo, but that'd be a bit too easy. So, instead, I run two Table of Joy sums in my head – just like the ones from the previous section. Here's what I do:

1. **Find the price per kilo of the first box of cereal.**

   Use the Table of Joy sum from the 'Buying by weight' section above.

2. **Find the price per kilo of the second box of cereal in the same way.**

3. **Compare the two prices to see which cereal has the lower price per kilo.**

Cooking by weight

You use maths in cookery to calculate how long you need to bake your concoction using a given formula. This is actually a lot easier than it looks. For example, if your recipe says 'Bake turkey for half an hour plus ten minutes per kilogram' and you know the weight of the turkey, you times the number of kilograms in the turkey by the time per kilogram, then add the extra time they mentioned first (in this case, half an hour). The numbers may change, as may the thing you want to cook, but the method stays the same.

You may need to do further sums with the time – for example, finding what time you'd expect the baking to finish, or what time you'd need to start cooking to be ready in time for your guests arriving. Look at the 'Doing Sums with Time' section of Chapter 10 to see how to do that.

(I pulled the numbers above out of thin air and do not recommend them as an actual turkey-cooking recipe. I have no experience with turkey cooking. This sum is for maths purposes only, okay?)

Chapter 13

Feeling the Heat

For years, I'd get a postcard every summer from my dad saying 'Having a great time here in [exotic place of choice]. Weather great, in the 80s most of the week. Wish you were here!' Don't get me wrong – my dad is one of the world's more creative people – but his postcard-writing is not one of the areas in which he exercises it.

But the temperature was always one of the most important things he could think of to mention about his holiday. 'In the 80s' is shorthand for 'between 80 and 90 degrees Fahrenheit', or 'between about 27 degrees Celsius and 32 degrees Celsius', or 'hot, but not unpleasantly so'.

You probably already have an understanding of temperature: the higher the temperature, the hotter it is and the more likely it is to burn you. Similarly, the colder something is, the lower the temperature and the more likely you are to freeze.

In this chapter I explain that Celsius and Fahrenheit are two different ways to measure temperature (you don't often need to convert between the two). I show you how to read a thermometer and use the concept of temperature to deal with ovens, weather and health. I help you recognise negative temperatures – they're the cold ones. I also explain how to find the difference between temperatures, including negative temperatures. Teachers at school often handle negative numbers very badly indeed, so I don't blame you if the thought of all those minuses makes you flinch. I work through negative numbers very slowly and carefully in this chapter, so I hope they make a bit more sense my way.

Understanding Temperature

Temperature is simply a measure of how hot or cold something is. In the UK, we tend to measure temperature in degrees Celsius, where 0°C is just cold enough to freeze water and 100°C is just hot enough to boil water. A kettle heats water up to 100°C (so the water boils), and a freezer cools things down to below 0°C (so they freeze).

A warm summer day is around 20–25°C, and a typical temperature for baking a pizza is about 200°C. As I found out in a soup-related disaster the other evening, a liquid at about 80°C is more than hot enough to scald you quite badly, so be careful.

TECHNICAL STUFF

Sniffing out different temperature scales . . . and armpits

Until quite recently, the UK weather report gave temperatures as two numbers, like this: 'Plymouth will see a high of 16 degrees Celsius. That's 61 degrees Fahrenheit.' These days, most people who use the Fahrenheit scale are Americans or postcard-writers – or possibly people taking your temperature.

In the 1720s, a universally recognised temperature scale didn't exist and there was no sensible way to measure temperature. So a Mr Fahrenheit defined 0 degrees as the temperature of the coldest liquid he could make (a mixture of ice, water and ammonium chloride) and 100 degrees as . . . go on, guess . . .

100 degrees Fahrenheit was the temperature of Mrs Fahrenheit's armpit. The scientists eventually chose something other than Mrs F's armpit to represent 100 degrees Fahrenheit, so that water

boils and freezes 180 degrees apart, but by then the damage was done.

About 15 years later, Mr Celsius did something only slightly less daft: he decided to make 100 degrees the freezing point of water and 0 degrees the boiling point. Mr Linnaeus waited for Mr Celsius to die and sneakily turned things around to what we know today: at 0°C water freezes and we get frost; at 100°C water boils.

A third scale, the Kelvin, is used almost exclusively by physicists. The Kelvin scale makes energy calculations simpler and is especially useful for the crazy cold temperatures some scientists are interested in: zero Kelvin is the same as –271.3°C, which is as cold as anything can possibly get. The general rule of thumb, throughout maths, is to use the unit that makes the sums easiest.

Fathoming Fahrenheit and Celsius

If you haven't read the sidebar, you've missed all the news from the early eighteenth century about Mrs Fahrenheit's armpit and Mr Celsius getting his temperatures the wrong way round.

The short and sweet version is that two scientists, Fahrenheit and Celsius, came up with different ways of measuring temperature. Just like some people prefer to use pounds and ounces to measure weight and others like kilograms and grams, so some people prefer to measure temperatures in Fahrenheit and others prefer Celsius.

Fahrenheit is the older of the two. In the UK, weather forecasters rarely give the temperature in Fahrenheit these days. Most of the world (except the US) now uses Celsius.

It would be pretty inconvenient (and probably rather creepy) to have to stick a thermometer into Mrs Fahrenheit's armpit every time you want to check a temperature in Fahrenheit. Happily, scientists have come up with a formula and a table to help you swap between Celsius and Fahrenheit, so you don't need to disturb the poor lady.

Using a formula

If you find formulas a bit frightening, you may want to look at the 'Magic Formula' section in Chapter 4 – I take you through this one gently as well.

Imagine your American friends have mentioned a temperature in Fahrenheit and you want to convert the temperature into Celsius. Start by calling the number of degrees Fahrenheit something sensible like F and then use the following formula:

$$C = (F - 32) \times 5 \div 9$$

Always work out brackets first – and then any powers, then multiplications and divisions, then additions and then subtractions. This formula tells you to take 32 away from the Fahrenheit temperature, times the result by 5, and then divide that result by 9. If my friend in Washington, DC tells me the temperature's 86°F over there, I take away 32 (to get 54), times by 5 (to get 270) and divide by 9 (to get 30). So 86°F is 30°C.

If you want to go the other way and convert Celsius to Fahrenheit, you use the formula in reverse, like this:

$$F = C \times 9 \div 5 + 32$$

If you're chilly at 5°C here in the UK, you do 5 times 9 (to get 45), divide by 5 (to get 9) and add 32 (to get 41). So 5°C is the same as 41°F.

Try practising the two formulas above by converting 41°F into Celsius (and make sure you get 5°C) and converting 30°C into Fahrenheit (and make sure you get 86°F).

Using a table or a scale

You can also use a table listing corresponding temperatures in Celsius and Fahrenheit, as I show in Figure 13-1. Here are the steps to follow:

1. **First check whether the information you need to convert is Celsius or Fahrenheit.**

2. **If you know the Celsius temperature, look at the Celsius column and find the temperature you want to convert.** If you know the temperature in Fahrenheit, look down the Fahrenheit column in the same way.

3. **The temperature next to it, in the other column, is the converted temperature.** This is your answer.

Celsius (°C)	Fahrenheit (°F)
−10	14
−5	23
0	32
5	41
10	50
15	59
20	68
25	77
30	86
35	95
40	104

Figure 13-1:
Converting temperature using a table.

In Figure 13-2 I show an example of a scale. You use scales in a similar way to tables, following these steps:

1. **First check whether the temperature you need to convert is in Celsius or Fahrenheit.** If you know the Celsius temperature, look at the side of the scale marked 'Celsius' or 'C'; if you know the Fahrenheit temperature, look at the side of the scale marked 'Fahrenheit' or 'F'.

2. **Find the temperature you want to convert on your chosen side of the scale.** If it isn't one of the numbers marked, you need to make an estimate about where it is – ask yourself if it's about halfway between two marks, or closer to one than the other. Make a small mark on the scale at that point.

3. **Read the value from the other scale at that point.** Again, your mark may not be exactly on a given number; in this case, you need to make a sensible estimate. The number you come up with is your answer!

Thermometers quite often have both Celsius and Fahrenheit marked on them, and you can use these scales to convert temperatures if you need to.

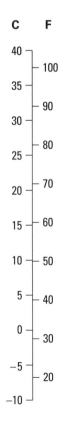

Figure 13-2:
Converting temperature using a scale.

Thinking about Thermometers

You measure temperature with a thermometer (*therm* means 'heat', and *meter* means 'measure'). You may encounter various types of thermometer.

The easiest to read is the digital thermometer, which displays the temperature as a number on a screen. The only dodgy thing to watch out for is a possible minus sign – check the section 'Nailing Negative Temperatures' later in this chapter for more on negative numbers.

Most other thermometers involve reading a scale of some sort. Most thermometers with scales have a straight-line scale, but some have circular scales. You work out the value of each tick or half-tick and read off the number. Head to the section 'Reading Scales' in Chapter 12 for more on reading a thermometer scale.

I give a couple of examples of thermometers showing 33°C in Figure 13-3.

Figure 13-3:
33°C on two types of thermometer.

When you read a scale, make sure the ticks mean what you think they mean. Here's the tick tactic:

1. **Find two labelled ticks and note how many degrees apart the two ticks are.**

   For example, in the second of the thermometers in Figure 13-3, find 30 and 40. They are 40 − 30 = 10 degrees Celsius apart.

2. **Count the number of ticks between the two chosen points.**

   In our example, we have ten ticks.

3. **Divide the degrees by the number of ticks.**

   In our example, we have 10 ÷ 10 = 1, so each tick is one degree.

4. **Count how many ticks above the first label the reading is and times this by how big each tick is.**

   In our example, we have three ticks times by one degree each, which gives three degrees.

5. **Add this to the first label to get your answer.**

   In our example, this gives 33°C.

If the thermometer reading isn't exactly on a tick, work out the temperature for the tick below and the tick above, and then make an intelligent guess about whether you're more or less than halfway between the two.

Looking at Everyday Temperatures

You use temperatures in three main areas of everyday life: setting the oven to bake at the right heat, checking the weather to decide how many layers to wear and whether the roads will be icy, and checking whether you have a fever when you feel ill.

Get cooking!

Ovens, really, are quite self-explanatory. If the package says 'Preheat the oven to 200°C,' you turn the little turny thing until it points to 200°C. Unless you're doing an experimental baked salad, this isn't rocket science.

In a maths test, you may need to convert a Fahrenheit temperature into Celsius for cooking using a given formula. Check out the section 'Fathoming Fahrenheit and Celsius' earlier in this chapter for more on how to convert temperatures.

Whatever the weather

You most probably see or hear about temperature most when you watch, read or listen to the weather forecast.

Your maths test may ask you to find the difference between the temperature in two different places, or between the highest and lowest (maximum and minimum) temperatures in the same place.

The test may also include questions about weather graphs and statistics – in this case, temperatures work just like any other kind of number. Head to Part 4, especially Chapter 16 on reading graphs and Chapter 18 on averages and ranges, for extra help on graphs and statistics.

Sploosh!

My local swimming pool always displays the water and air temperatures marked somewhere prominent. For some reason that I don't know, the water is always slightly warmer than the air, but not by much.

You may need to answer a test question on the difference between the swimming pool and air temperatures – or, how much warmer is the water than the air? Think of this difference as a simple take-away sum, which only gets a little tricky if the temperatures aren't whole numbers – in which case, check out Chapter 7 on decimals.

Fever!

As a child I hated my mum taking my temperature when I was ill. I suspect it may have been something to do with my mum, who used to be a nurse, taking the thermometer out of my mouth, reading it and shaking it *without letting me see*. How was I supposed to know how ill to behave if I didn't know how hot I was?

To take someone's temperature, you put a medical thermometer under their tongue and wait for a certain time, according to the instructions that came with the thermometer. Then you take the thermometer out and read the scale or the digital display. Medical thermometers are quite small, so the numbers on the scale can be hard to read. However, most modern medical thermometers are digital, so you simply read the number off the screen.

Don't use a house or garden thermometer to take a person's temperature: those bad boys are full of mercury – if you're not already sick, mercury will probably take care of that.

The average human body temperature is around 37°C. Older sources of information often give the average temperature in Fahrenheit as 98.6°F, but that's inappropriately precise. As a general rule, if the temperature under your tongue is significantly higher than 37°C, you probably have a fever.

Nailing Negative Temperatures

Unless you live somewhere tropical (my geography isn't great, but as far as I know not much of the UK is anywhere near the equator), you have probably experienced sub-zero temperatures at some point. When the air temperature drops below zero, we measure the temperature using *negative numbers* – normal numbers with a minus sign in front of them, such as –2°C. The negative number shows how many degrees below freezing the temperature is.

Normal numbers, without the minus sign, are known as *positive* numbers. Unless I'm on a skiing holiday, warm temperatures are generally good news and make me feel positive, and cold temperatures are a pain in the neck (for me, at least) and make me feel negative.

Negative temperatures can also make people grumpy because the numbers seem to go the wrong way. A temperature of –22°C looks like it ought to be warmer than a temperature of –15°C, because 22 is bigger than 15. In fact, –22°C is 22 degrees colder than freezing, and –15°C is only 15 degrees colder than freezing, so –22°C wins in the 'wrap up extra snug' stakes.

The concept of cold temperatures and negative numbers is one of the few arguments in favour of the Fahrenheit scale: most of the time in Fahrenheit, we don't worry about negative numbers, because 0°F is about –17°C, a temperature most of Britain very rarely sees.

For practical purposes, we care about negative temperatures in real life mainly for weather reports and for using the freezer.

Ordering negative temperatures

Putting temperatures in order, from coldest to hottest, or vice versa, is fairly straightforward, even when we have some negative numbers. Negative temperatures are *always* colder than positive (normal) temperatures, and the bigger the number after the minus sign, the colder the temperature is. Try following my recipe below, in conjunction with Figure 13-4, to put a group of temperatures in order:

1. **Circle all of the negative numbers so you don't mix them up with the positive numbers.**

2. **Find the most extreme circled number – the one furthest from zero.** This would be the biggest circled number if you ignored the minus sign. Label this number '1'.

3. **Find the biggest number you haven't labelled yet and put the next label ('2', '3', and so on) by it.**

   Repeat Step 3 until you've dealt with all the circled numbers.

4. **Find the smallest uncircled number and put the next label next to it.**

5. **Find the smallest uncircled number you haven't labelled yet and keep on labelling in this way until you run out of numbers to label.**

6. **Write down the temperatures in the order you've labelled them – they should now be in order from cold to hot.** (To order them from hot to cold, simply reverse the list.)

12	0	−10	27	−5

Figure 13-4:
Ordering temperatures.

12 (4) 0 (3) (−10) (1) 27 (5) (−5) (2)

In order: −10, −5, 0, 12, 27

Finding the difference between negative temperatures

In this section we get into the dreaded 'minus number' maths. I remember minus numbers being a real struggle in school, for myself and everyone else, until we started to think of minus numbers as points on the number line, like the one I show you at the beginning of Chapter 3. Here are some places you might need to figure out the difference between temperatures:

> ✔ Comparing maximum and minimum air temperatures from a weather report.
>
> ✔ Deciding how much warmer it is in one place than another.
>
> ✔ Seeing how much warmer a defrosted freezer is than a working freezer.

You probably already know how to find the difference between two positive temperatures – you do a simple take-away sum: the difference between 24°C and 16°C is 24 – 16 = 8 degrees.

With two negative numbers, the drill is almost exactly the same: ignore the minus signs and take one number away from the other. The difference between –9°C and –3°C is simply 9 – 3 = 6 degrees.

The numbers way

If you have one negative number and one positive number, things get interesting. The number in the negative temperature tells you how many degrees the temperature is below zero, while the positive number is how many degrees above zero the temperature is. To get from the negative number to the positive number, you have to *increase* by how many negative degrees you have, and then *increase* again by the number of positive degrees.

Instead of doing a take-away sum, you end up doing an adding sum. So when the signs are opposite, you do the opposite of what you're asked!

Here's my recipe to find the difference between a positive and a negative temperature:

1. **Drop the minus sign from the negative temperature.**

2. **Add the two numbers together.**

    This gives the temperature difference.

If you started with a cold, negative temperature and now have a warm, positive temperature, the temperature's gone up.

And if you started with a warm, positive temperature and now have a cold, negative temperature, the temperature's gone down.

Your maths test may contain a question like 'The temperature in the Sahara Desert drops from 45°C in the daytime to –10°C at night. What is the difference between those temperatures?'

Following the steps from my recipe, you drop the minus sign, to leave the numbers 45 and 10. Adding those up gives 55°, which is the difference between day and night in the desert.

The number-line way

You can look at differences between positive and negative temperatures using a number line. Have a look at Figure 13-5 and try following these steps to use the number line:

1. **Draw a line and mark a zero somewhere near the middle.**

   You only need to draw a rough-and-ready number line, so don't worry about measuring anything.

2. **Mark your two temperatures in the appropriate places.**

   Put positive numbers on the right and negative numbers on the left. Label the two numbers.

3. **Write down how far below zero the negative number is.**

   For instance, −10°C is 10 below 0. Write this description between the marks for the number and zero, as in Figure 13-5.

4. **Write how far above zero the positive number is.**

   For instance, 45°C is 45 above 0. Again, write this description between the marks for the number and zero, as in Figure 13-5.

5. **To get from the negative number to the positive one, you have to move right by the *total* of the numbers – so add up your two numbers.**

   This is your answer.

Figure 13-5:
The difference between day and night: negative temperatures and the number line.

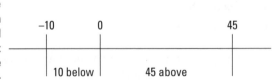

Total: 55 degrees difference!

The difference between two temperatures is also called the *range*. Your maths test may include a question such as 'What is the temperature range?' I talk about ranges in more depth in Chapter 18.

Temperature ranges from a table

A particular favourite of examiners is to give you a table of temperatures and ask you to find the difference between the warmest and coldest temperatures. This adds an extra layer of complexity, but don't panic. Here's what you need to do:

1. **Find the warmest temperature – the biggest number without a minus sign. If they all have minus signs, pick the one closest to zero.**

   Circle this number or write it down.

2. **Find the coldest temperature – the number with a minus sign that's furthest from zero; if none of them have minus signs, pick the smallest number.**

   Circle this number or write it down.

3. **If both numbers are positive, or both are negative, take away the small number from the big number as normal.**

4. **If one number is positive and one number is negative, ignore the minus sign and add the two numbers.**

   Look back to 'The numbers way' earlier in this section if that sounds a bit fishy.

Don't overcomplicate things! Always try to find the simplest way to do your temperature sums, and don't be afraid to draw out a number line to help you.

Chapter 14

That's About the Size of It

*I*n this chapter, I introduce the idea of size in all its forms – the different ways you can measure how big an object is. Whether something is big or not is largely a matter of context, just like when you talk about a number being a lot. Is 12 a lot? Not if you're talking about grains of rice, but if you're talking about children . . .

Rather than just saying 'That's a big house,' you can put a number on exactly how big the building is. In fact, you can use several numbers: you can talk about how far the walk around the house is (the perimeter), and how many metres tall the house is (the height), and how wide or far back the building goes (the width and depth). These are all one-dimensional ideas – you can measure them directly with a ruler, and you can report your answer in metres or other units of length.

In this chapter I guide you through measuring distances and switching back and forth between different units of measurement.

You can also describe the size of the house by talking about its floor space – or the area of floor available to use. This is a two-dimensional idea – and difficult to measure directly. In this chapter I show you how to measure the area of spaces. Fortunately, most houses are nice simple shapes and therefore house areas are quite easy to figure out. We can also talk about the capacity or volume of the house. These are three-dimensional concepts – but don't let that description worry you: most of the time capacity and volume are easy to work out, as I show you in this chapter.

How Big Is That Suitcase?

I start this section with a controversial statement: big is a meaningless word.

For example, is *Basic Maths For Dummies* a big book? Yes, compared with *A Little Algebra Book*. But compared with *Encyclopaedia Britannica*, my book's a tiny pamphlet. And compared with another For Dummies book – this one's about average.

Fortunately, we have several ways to measure bigness. Some of these methods involve the *linear* size of objects – how wide or tall or deep something is. Some methods involve *cross-sectional area* – for example, can a letter fit through this slot? Other methods involve the *volume* – how much space something takes up.

Checking in: Dimensions of luggage

If you've flown on a plane recently, you may know that some airlines have baskets for you to check whether your carry-on bags are too big for the cabin.

Written above these baskets are the dimensions of luggage you're allowed to take on board. The exact size depends on the airline you fly with, but typically the size is about 55 cm by 40 cm by 20 cm – which is just a fancy way of saying your bag needs to fit into a basket 55 cm deep, 40 cm wide and 20 cm across.

These are all measurements of *length* or *distance*. You can measure each of them with a ruler. The basket acts as three rulers at once – it measures the width, the height and the depth of your luggage.

Using a basket isn't the only way to measure your luggage.

The airline may also be interested in the *area* of the base of your suitcase – will your bag fit nicely on the floor of the hold of the aircraft? – and the *volume* – how much space will your bag take up?

The volume of a shape has nothing to do with how loud the thing is.

Sizing up the vocabulary you need

Most of the words you use to describe size are common English words. You probably know most of these words already, but I describe them here just in case: after all, sometimes the maths version of a word is a bit more strictly defined than the normal version:

- **Area:** How much paper or something similar you need to cover a surface. We use different formulas to work out area for different shapes. We measure area in centimetres squared (cm^2), or metres squared (m^2) or any other unit of length squared.

- **Capacity:** How much stuff can fit in an object. We usually measure capacity in litres, but sometimes we use centimetres cubed (cm^3) or metres cubed (m^3).

- **Circumference:** The perimeter of a circle. To work out the circumference of a circle, we use the following formula: two times pi times the radius of the circle, or $2\pi r$. Pi (written as the Greek letter π) is a number a little bigger than 3.14.

- **Distance:** How far one thing is from another thing. We measure distance in units of length, such as centimetres, metres, feet or inches.

- **Length:** How long an object is. We normally measure length with a ruler in centimetres, metres, feet or inches.

- **Perimeter:** How far around a shape is – that is, if you walked all the way around a shape, how far you would travel. We measure perimeter in units of length, such as centimetres, metres, feet or inches. You find the answer by adding up the length of all of a shape's sides.

- **Volume:** How much space an object takes up. We measure volume in centimetres cubed (cm^3) or metres cubed (m^3) or any other unit of length cubed.

Mathematicians and scientists often describe mathematical shapes as 'thin', which means the amount of space they take up and the amount of stuff they contain is the same – so we can consider volume and capacity as the same thing.

Meeting Some Common Measuring Tools

The most common measuring tools you come across are straight edges with distances marked on them – rulers and tape measures are obvious examples.

You may also see callipers (which look a bit like a pair of compasses), trundle wheels (which seem to have gone completely out of fashion) and those laser-type things workers in yellow jackets use. (I keep promising myself to ask if I can have a go the next time I see them out surveying.)

Reading a ruler

In this section I give you a quick rundown on using a ruler. Apologies if you think working with a ruler is a basic skill, but you may be surprised to know that many perfectly intelligent people mess up making even the simplest measurements. For example, I once missed out on full marks in an exam because I mismeasured a circle. The shame has haunted me through my career, and I want to save you from that particular horror.

Having thoroughly revised the use of rulers, I'm now in a position to tell you how to measure a distance:

1. **Decide which unit you want to use.**

   Many rulers have inches on one side and centimetres on the other side. Choose the side you need to work in.

2. **Find the zero on the side of the ruler you want to use.**

   The zero is normally very close to the left-hand end.

3. **Put the zero mark over the start of the thing you want to measure.**

   Line up the ruler so it goes to the other end of the thing you want to measure.

4. **Check the zero mark is still where you want it to be.**

   Read the number on the ruler where the thing you're measuring ends. The number is your length.

In Figure 14-1 I show a ruler in action. Try to contain your excitement.

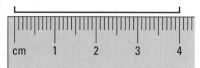

Figure 14-1: The line is 4cm long.

Minding the marks

A common problem in measuring – whether you use scales, a thermometer, a protractor, a graph, or a dozen other things – is how to deal with a reading that falls between two marked numbers. In this section, I assume you're using a ruler, but my tips apply to any measuring device with a scale on it.

As I show in Figure 14-2, your ruler may have little marks between the numbered marks – or it may not.

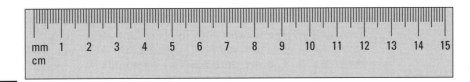

Figure 14-2:
A ruler with little marks, and a ruler without.

The way you measure changes slightly depending on the kind of ruler you have: the ruler with the small marks allows you to be a little more accurate and to use a little less guesswork. Here's what you do when your measurement lies between two marked numbers and you have the little marks:

1. **Count the number of little spaces between your marked numbers.**

   This number's normally ten, but some rulers are sneaky. Some rulers divide inches into sixteenths, which is almost perverse.

2. **Take your main unit – however far apart your marked numbers are (usually 1 cm) – and divide it by the number in Step 1.**

   With a metric ruler, your answer is generally a nice decimal, but you may need a nastier fraction with a non-metric ruler.

3. **Count how many little marks past a marked number your measurement is.**

   If you're between two little marks, I salute your eyesight and attention to detail! Just pick whichever is closer.

4. **Work out how far above a marked number you are by timesing your answers from Steps 2 and 3.**

   If your marks are 0.1 cm and you're six marks above the marked number, the number is 0.6 cm.

5. Add this to the marked number below.

Using the example from Step 4, if you're between 5 cm and 6 cm on the ruler, your answer is 5.6 cm.

Often the halfway mark between two centimetre marks is bigger than the other little marks. Try using the halfway mark to make sure you've counted correctly.

If you don't have little marks between the marked numbers, you need to use some intelligent guesswork. You make the following series of decisions:

1. Is your reading more or less than halfway between the two marks?

If more than, your answer ends in something bigger than '.5'. If less than, your answer ends in something smaller than '.5'. If your reading's just about exactly halfway, your measurement ends in '.5'.

2. Is your measurement closer to halfway, or closer to one of the marks?

Use this question to decide between '.6 and .7' and '.8 and .9' for the last digit if you're more than halfway, or between '.1 and .2' and '.3 or .4' if you're less than halfway.

3. Make a choice between your two remaining options, largely based on gut feeling and eyesight.

4. Tack your decimal digit on to the lower number of the two marks.

Using Different Units of Length

A ruler often has a different scale on each of the two sides: one scale may go from 0 to 30 or 0 to 15 in spaces each measuring a centimetre – about the width of a finger – and the other scale may go from 0 to 12 or from 0 to 6 in inches – about the width of two fingers (all depending on the size of your fingers, obviously).

A hundred centimetres make a metre, as the name implies, and centimetres are part of the metric system – the same system that uses kilograms and grams for mass, which I talk about in Chapter 12. The idea of the metric system is that converting between units is simply a matter of timesing or dividing by 10, 100 or 1,000.

Inches, on the other hand, are part of the more complex, traditional, imperial system. Twelve inches make a foot. Three feet make a yard. A furlong contains 220 yards. Eight furlongs make a mile . . . maybe you see why using the metric system can be easier than using the imperial system.

Starting off on the right foot

A *foot* was originally just that – about the size of a normal adult foot. A grown man's foot isn't a very practical thing to use if you want to measure something accurately, but describing someone as six feet tall would give you an idea of how tall the person was. At some point in the history of measuring things, somebody somewhere decided to standardise the foot. We now define one foot as 30.48 cm – which is why we use rulers that are 30 cm long.

One inch is about 2.5 centimetres long – or more precisely, 2.54 centimetres. A yard is 3 feet, or 36 inches or 91.44 centimetres, or a little less than a metre. A mile is 1,760 yards or a little over 1,600 metres. You may see furlongs (220 yards) used in horse racing, but pretty much nowhere else.

To convert from inches into centimetres, you just multiply by 2.54. But the tricky bit is working out how many inches you have to start with. Here's a step-by-step guide:

1. **Look at how many feet there are and times that number by 12.**

   Remember: 1 foot is 12 inches. For example, if you're 5 foot, 7 inches, do $5 \times 12 = 60$.

2. **Add on how many 'spare' inches you have.**

   In our example, we have seven 'spare' inches. $60 + 7 = 67$.

3. **Times the result by 2.54.**

   For our example, do $67 \times 2.54 = 170.18$ cm.

 I shouldn't give the result that precisely: 5 feet, 7 inches is a pretty rough measure of height and I certainly can't justify an accuracy to the nearest hundredth of a centimetre. A more sensible answer is 'about 170 cm'.

Looking at Length, Distance and Perimeter

Length, distance and perimeter are all measures of length. They all behave in much the same way, and we measure all three in units of distance, such as metres, centimetres, kilometres or similar imperial units. The difference between the three is really one of language and application, rather than how they work.

How long is a piece of string?

The *length* of an object is how far apart the two ends of the object are. To find the length of a piece of string, you might put one end of the string on the zero mark of a ruler, pull the string straight and read off the ruler where the string ends. Check out the section 'Meeting Some Common Measuring Tools' for more on using a ruler.

A length is normally the answer to the question 'How long is [one thing]?'

How many miles to Babylon?

A *distance* is how far apart two things are. My house is 250 metres away from the railway station and New York is 4,500 kilometres away from Los Angeles.

A distance is the answer to the question 'How far apart are [two things]?'

The difference between length and distance is subtle and often barely exists. If the shortest distance between two towns is ten miles, then the length of the straight road between the towns is also ten miles. Generally, when you talk about the size of a single thing, you call it a length; when you talk about how far apart two things are, you call it a distance.

Going all the way round

Airports and military bases have *perimeter fences*, some of which are occasionally scaled by protesters and undercover police officers. These fences are so-called because they go all the way around the site – which is what the perimeter measures. (*Peri-* means 'around', as in 'periscope' – a contraption for looking around things.)

The *perimeter* of a thing is how far it is all the way round the thing. You find the perimeter of a rectangle by finding the length of each of the four sides and adding up the four lengths. You could also run a piece of string all the way round the rectangle and then measure the amount of string you use – but this is a fiddly way of doing an easy job.

A perimeter is the answer to the question 'How far is it round [a thing]?'

Summing up distance

Doing sums with distance is similar to doing sums with any other kind of measure. If you walk 10 metres and then another 20 metres, you walk a total of 30 metres – you just add up 10 + 20 as normal. If you walk 50 metres, realise you dropped your wallet 20 metres back, and turn around to fetch the wallet, you end up 30 metres from where you started – you just take away 50 – 20 = 30.

Lengths and distances add up as normal as long as you travel in the same straight line. If you walk 10 metres in one direction and then 10 metres in another, however, you could end up anywhere from 0 to 20 metres from where you started.

You can also times and divide distances by numbers. For example, if you have five 30 cm rulers and place them end to end, their total length is 5 × 30 = 150 cm. If you snap one ruler in half (don't try this at home, kids), each part is 30 ÷ 2 = 15 cm long.

Accessing All Areas

The *area* of a shape is how many squares of paper (of a given size) you need to cover the shape. Everything else being equal, the bigger the area of a meadow, the longer you need to mow the grass. We often measure smallish shapes, such as pieces of paper, in square centimetres (cm^2) or square millimetres (mm^2); medium-sized shapes, such as rooms and parks, in square metres (m^2); and huge shapes, such as countries, in square kilometres (km^2).

If you're feeling imperial, you can use square inches, square feet and square miles.

To understand what the 'square' in 'square centimetre' means, imagine a square where all of the sides are 1 cm long, as in Figure 14-3. The area of this square is defined as one square centimetre. We use a similar idea for square metres and square miles: if you draw a square (maybe in chalk in a car park) with sides measuring one metre each, the area is one square metre.

In Figure 14-3 I also show that you can think of the area of a shape as the number of little squares you need to cover the shape. In the next section I help you work out areas without counting.

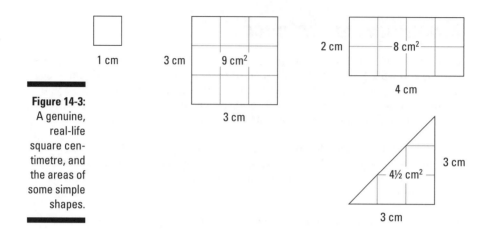

Figure 14-3:
A genuine,
real-life
square cen-
timetre, and
the areas of
some simple
shapes.

Although 1 metre is the same as 100 centimetres, 1 square metre isn't the same as 100 square centimetres. If you draw a square metre and put a row of 100 square centimetres inside it, you cover only one edge of the square. You actually need another 99 rows of square centimetres to fill up the square. One square metre is actually $(100 \text{ cm}) \times (100 \text{ cm}) = 10{,}000 \text{ cm}^2$.

Recognising rectangles

In Figure 14-3, you can see just by counting that the rectangle in the middle – 2 cm tall and 4 cm wide – has an area of 8 cm^2. You may notice that if you have two rows, each with four boxes, you end up with $2 \times 4 = 8$ boxes. Equally, you may see this as four columns with two boxes each – both methods give you the same answer. These rules are true for all rectangles. If you have a box, 12 cm wide and 5 cm tall, its area is $12 \text{ cm} \times 5 \text{ cm} = 60 \text{ cm}^2$. This even works if the lengths of the sides aren't whole numbers: you just times the numbers together and whatever comes out is the area.

Be very careful with units when you work out areas. Occasionally an examiner sneakily measures one side in centimetres and another side in millimetres. Before you do the sum, either convert the millimetres into centimetres or vice versa.

Joining things up: Compound rectangles

One of the most troublesome things you have to do with areas is find the total area of a shape made up of several rectangles. This may seem tricky, but after you see what's going on, things start to make sense.

The trick is to split up the shape into smaller rectangles for which you either know or can figure out the length of each side. This is a little bit more of an art than a science – but as with everything else in maths and life, the more you practise, the easier the method gets.

Take the following steps if you need to find the area of a shape that isn't a rectangle but is full of right angles:

1. **Try to find somewhere to draw a line that splits the shape into two, smaller rectangles.**

   If not, draw a line that splits off one rectangle and try to split the rest of the shape up into smaller rectangles. How many rectangles you have doesn't matter – but you're less likely to make mistakes if you use as few shapes as possible.

2. **Work out the sides of each rectangle you have left over.**

   You may need to do a bit of lateral thinking, but normally you just need to take something away from the total length of a side.

3. **Find the area of each rectangle by multiplying the sides together.**

   Write the answer in the middle of the rectangle so you don't forget it.

4. **Add up all of the areas you just worked out.**

   The answer is your total area.

In Figure 14-4 I show an example of this method.

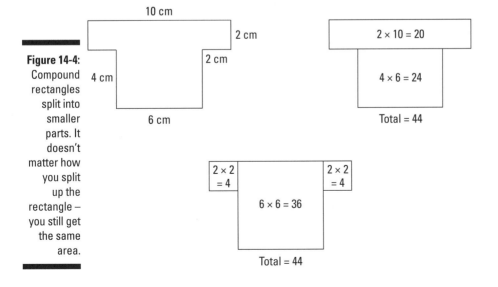

Figure 14-4: Compound rectangles split into smaller parts. It doesn't matter how you split up the rectangle – you still get the same area.

In Figure 14-4, I start with a wibbly shape and neatly cut it into two rectangles with a sideways line near the top. I work out the height of the bigger rectangle, which is the only side I didn't know to start with. I find the areas of the two rectangles and add up these two areas to get a final answer.

On the right of Figure 14-4, I split up the rectangle in a different way, giving three separate shapes. Again, I work out the lengths of the sides, find the areas and add the areas to get a final answer. I end up with the same answer using both methods.

If you want to feel very proud of your estimation skills, start by thinking 'If I imagine a rectangle as tall as the total height of the shape, and as wide as the total width, what is the area of the shape?' Then say 'My total area has to be smaller than that.' This estimate gives you a quick-and-dirty check on whether your eventual answer is correct, and in multiple-choice tests can help you eliminate one or two of the answers.

Using formulas

Occasionally in a maths test you may have to use a formula to work out an area or volume. Formulas look a bit intimidating, but if you stay calm and think logically you can do it.

If you're unfamiliar with formulas, read the section 'Figuring Out Formulas' in Chapter 4 to discover they're less scary than you think.

Each letter in a formula represents a number. You need to work out what letter goes with what number, write out the sum and figure out the answer.

Here's how I deal with formulas of any variety, and particularly area formulas:

1. **For each letter you know, write down the letter next to what it equals.**

2. **Go through the formula one symbol at a time and turn it into a sum by replacing each letter with its number.**

   If you have two letters next to each other with no symbol between them, write a times sign between them.

3. **Do the sum, being very careful about the order of doing things.**

   If there's a little two above anything, times that by itself first. Then do all of the timeses and divides (working left to right), and then all of the adds and take aways (again, from left to right).

Here's an example of using a formula to work out an area:

The surface area of a pyramid is $x^2 + 2xL$. The width of the base is x. The sloping distance up the middle of a side is L. The Great Pyramid at Giza is about 200m wide and has a sloping distance of about 220m. What is its surface area?

1. **My formula is $x^2 + 2xL$, and the question says that x is 200 and L is 220.**

   That means I have to work out 200 squared and add on 2 times 200 times 220.

2. **200 squared is $200 \times 200 = 40{,}000$.**

3. **$2 \times 200 \times 220 = 400 \times 220 = 88{,}000$.**

4. **$88{,}000 + 40{,}000 = 128{,}000$.**

   The surface area is $128{,}000\text{m}^2$.

See the sidebar 'Zeroing in . . . and out again' if all those zeros confuse you.

Verifying Volume and Capacity

The *volume* of an object is how much space the object takes up – or, if you were to drop the object into a full tub of water, how much water would overflow. *Capacity* is how much space an object has inside – or, how much water you can fit inside the object. This distinction between volume and capacity is subtle – we can measure both in cm^3, although confusingly we can also measure capacity in millilitres (ml), each of which is the same size as 1 cm^3. A litre contains 1,000 millilitres, and a cubic metre contains 1,000 litres.

Incidentally, a cubic centimetre is the volume of a cube which has edges that are one centimetre long – about the size of a normal die.

For the numeracy curriculum, you may need to work out the volume of a *cuboid* or shoebox. You normally know the width, height and depth of the box. To work out the volume, you simply times the three numbers together.

A classic problem in numeracy exams involves working out how many small boxes fit into a bigger box. This kind of packing problem has real-life applications (how many DVDs can you fit into a box? Will this crate hold all the copies of *Basic Maths For Dummies* you want to send to your friends around the world?) and is quite straightforward. In an exam, you normally know the *orientation* – or which way round you need to pack the little boxes into the big box. Look at Figure 14-5 and follow these steps to work out how to fit little boxes into a bigger box:

1. **Work out how many boxes you can fit along the front of the box.**

   Divide the width of the big box by the width of one small box and write down the result. If you get a whole number answer, great! If not, round *down*, because even if your answer is 5.99, you can't squeeze a sixth little box into the crate.

2. **Work out how many boxes you can fit along the side of the box.**

   Divide the depth of the big box by the depth of the little box and write down the answer. Round down if you don't have a whole number.

3. **Work out how many boxes you can fit going up the box.**

   Divide the height of the big box by the height of the small box and write down the number. Round down if you need to.

4. **Times the three numbers together.**

   That's your answer!

Here's a typical question to follow as an example:

> *A crate is 4 m wide, 12 m long and 3 m deep. You want to fill it with boxes that are 2 m wide, 3 m long and 1 m deep. How many boxes will fit in the crate?*

1. **You can fit two boxes along the width of the crate.**

2. **You can fit four boxes along the length of the crate.**

3. **You can fit three boxes along the depth of the crate.**

4. **You need to times those numbers together.**

   $2 \times 4 \times 3 = 8 \times 3 = 24$.

5. **You can fit 24 boxes into the crate.**

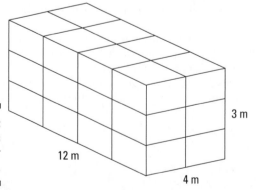

Figure 14-5: Fitting boxes in a bigger box.

3 m

12 m

4 m

Reading Maps and Plans

Being able to read maps may be a less vital skill than it used to be: the rise of the GPS and satnav has almost eliminated the need for motorists to know the difference between a church with and without a steeple. But map-reading is still a useful ability to have, especially if you take part in outdoor activities such as hiking or mountain-biking: your ability to work out which route to take can make the difference between people in your group loving or hating you; or between living and dying. Nobody's going to ask you to draw maps or even read a map in great detail at this stage of your maths career. But you do need to be able to look at a map and understand what it says, to convert distances using the scale (this is the one that comes up most often in tests), and to understand how to see a plan simultaneously as a 'real' object and a mathematical shape.

A map is a geographically accurate picture of a piece of land, highlighting some features, such as roads, rivers, buildings, terrain and hills, but ignoring others, including things that move, small things and irrelevant things. Different types of map highlight different features – so a walking map tells you much more about the terrain than a driving map. Each feature on a map normally has its own symbol. Roads may be different-coloured lines, pubs may be marked PH, and towns may marked with their names.

As far as I can tell, for the numeracy curriculum you mainly need to know about road maps. For example, you may need to find the shortest route between two towns along two different roads – you find the distance of each route and see which is shorter.

Scales and distance

One of the most important things to know about a map is its *scale*. This ratio tells you the relationship between distances depicted on the map and distances in reality.

If a map has a scale of 1:100,000, it means that 1 centimetre on the map represents 100,000 centimetres (the same as 1 kilometre) in real life. The scale allows you to say how big a real-life object is if you know how big its picture is on the map, as well as how big something should be on the map if you know its size in real life.

The most common map-based questions involve solving this type of problem. The exam question may show you a map and ask you to do some measuring, or the question may just give you the scale and a distance (either on the map or in real life) and ask you to work out the other distance.

Scales questions are fiddly and often involve big numbers and conversions. Slipping up is very easy, so take extra care to work neatly so you can see what you're doing. Write down your units as well – mixing up kilometres and centimetres can give you wildly wrong results!

A scale looks like a ratio – two numbers with a colon in between. The map is always smaller than real life (you don't ever see a map of Scotland bigger than Scotland itself). The smaller number in the scale always refers to the map. For example, a scale of 1:10,000 means that 1 cm on the map represents 10,000 cm in real life – which is 100 m. You can do ratio calculations with scales using the Table of Joy, which I describe in Chapter 8, but unless you have a scale that's not 'one to some big number', the Table of Joy is slightly overkill. Instead – as long as you can keep your units straight – you can get by with timesing or dividing. If you have a stranger scale such as 2 cm to 5 km, you may do better using the Table of Joy.

Scales with a '1:'

The key to doing scales sums without the Table of Joy is to remember that multiplying by numbers bigger than one makes things bigger and dividing by numbers bigger than one makes things smaller. Here are the steps to do a sum involving a 'one to lots' scale, starting from a distance on the map:

1. **If you don't know the distance on the map, measure the distance on the map.**

   Write down the distance you measure, giving the right units – normally centimetres.

2. **Times the map distance by the scale.**

   Be very careful to use the right number of zeros. I always lose count of noughts!

3. **Convert your big number into more appropriate units.**

   If you're working in centimetres, divide by 100 to get metres. If you need an answer in kilometres, divide the number of metres by 1,000. The number is your answer.

Scales without a '1:'

Sometimes you see a scale that I describe, with a complete lack of affection, as 'silly' – say, something like '2 centimetres to 5 kilometres'. The only redeeming features of such monstrosities are that they tend to use whole numbers and you can use the Table of Joy on them easily.

Zeroing in . . . and out again

If your number has lots of zeros – say, 1,000,000 – keeping track of your sums is really tricky. Plus copying long lists of zeros is so tedious.

Coming up with a strategy for dealing with lots of zeros makes sense.

My favourite way to deal with many zeros is to forget about those noughts for a while. When I times numbers with lots of zeros, I write down how many zeros I have and do the sum as if the zeros didn't exist – then I throw them back in at the end.

If you deal with different units at the same time – kilometres, metres and centimetres, for example – first work on the number of zeros. Say you have five zeros on your number of centimetres: you know 100 cm makes a metre, so your number of metres has two fewer zeros. To report your answer in metres, you need to use three rather than five zeros. You can go further too: you know 1,000 m makes a kilometre, so you can knock off another three zeros – which in this example means you end up with no zeros at all.

Here's how to throw the Table of Joy at this kind of scale and get an answer without too much effort:

1. **Draw a Table of Joy noughts-and-crosses grid.**

   Leave plenty of space for labels.

2. **Label the columns 'map cm' and 'real life km'.**

   Label the rows 'scale' and 'measured'.

3. **Fill in the scale row according to the scale on your map.**

   Fill in any other information you have: if you know a real-life distance, put it in the real-life column; if you have a map distance, put it in the map column.

4. **Put a question mark in the remaining cell and write down the Table-of-Joy sum.**

   Do the other number in the same row times the other number in the same column, divided by the number you haven't used yet.

5. **Work out the sum.**

   That's your answer.

The best-laid plans

A slight variant on a map is a plan. The only real difference is one of size: a map tends to show how to navigate fairly large areas – say, building size and up – and a plan shows the general layout of something smaller – maybe a room.

You can often deal with plans in the same way you play with compound rectangles, which I describe in the section 'Joining things up: Compound rectangles' earlier in this chapter. For example, you may need to find the perimeter or the area of the shape of the plan.

When you work with a scale, always convert your measurements from map distances to real-life distances before you start doing the shape sums – especially when you work with area.

Working out perimeter when you have a scale

To work out the perimeter of a shape in a scaled plan, follow these steps:

1. **If you don't know the lengths of the sides of the shapes, measure them or work them out based on what you do know.**

   Make sure you know all of the lengths on the edge of the shape.

2. **Scale each length up to the real-life length.**

3. **Add them all up.**

   The answer is your perimeter.

Calculating area when you have a scale

To work out the area when you have a scale, you first convert all of your lengths into their real-life sizes. Then the job's pretty much the same as working out an area as in the 'Accessing All Areas' section earlier in this chapter. Here are the steps you need to take:

1. **If you can't easily work out the area of the shape, split it into smaller, neater shapes. If the shape is a nice simple rectangle or a square, you can just skip straight to Step 2.**

   You can normally get away with two or three rectangles – the fewer the better.

2. **Find the width and length of each smaller rectangle by measuring with a ruler.**

   If the measurements are already marked, you don't need to measure.

3. **Convert your lengths in Step 2 to real-life measurements.**

   Use the Table of Joy or another method that works and you like.

4. **Work out the real-life area of each shape.**

   If your shapes are rectangles, do height times width for each rectangle.

5. **Add up all of your real-life areas.**

   The total is your area.

Interpreting plans in a maths exam

A plan is supposed to be a way of simplifying complicated shapes so you can understand the layout of, say, a room. Unfortunately, this description doesn't seem to have reached maths examiners, so they persist in giving you plans that look like an explosion at the bingo hall – numbers everywhere and a load of annoyed people trying not to trip up.

In an exam question, you probably won't need to measure the plan but you do need to figure out what the question means. Here are a few pointers:

- ✔ **Width and length:** If you need to find the width or length of a shape, check whether the plan already shows the measurement – if you see an arrow that goes all the way across or all the way along the shape, the number on that arrow is your width or length. If not, you probably need to add up the component parts: find a series of arrows that go all the way across or along the shape and add them up.

- ✔ **Perimeter:** To find the perimeter, you need to add up all of the lengths around the edge of the shape. If any of the lengths are missing, find them by taking away the appropriate arrows from the full length.

   If the perimeter has gaps, like a fence with gates, you can either leave the gaps out as you add up, or work out the whole perimeter and then take away the gaps at the end.

- ✔ **Area:** To find the area of a plan, split up the shape into simpler shapes and add up the area of each shape. The total area is the sum of all the other areas. Make sure you use the same units throughout.

- ✔ **Area of the walls:** To work out the area of the walls, you need to work out the perimeter and times this by the height of the walls. If the walls contain doors or windows, you then need to take the area of these away from the total wall area.

- ✔ **Volume:** To find a volume, such as how much water you can fit in a swimming pool, you find the area and then times it by the depth of the pool.

In particularly brutal exams, the examiner may ask you to do further work, such as how many buckets of water you need to fill the swimming pool or how many tins of paint you need to cover the walls.

If this happens, don't panic: you just divide the volume of the pool by the capacity of the bucket, or the area of the walls by the area each tin of paint covers.

TIP

Before you work with a plan, take a moment to think about how you'd answer the question if it wasn't anything to do with a picture of a shape. Ask yourself, for example, 'If I have a volume to fill and each bucket holds this much water, how do I figure out the number of buckets I need?'

Plan of a pool

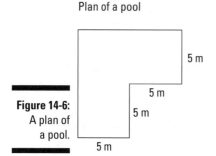

Figure 14-6:
A plan of
a pool.

In Figure 14-6 I show an example of a plan of an L-shaped pool. I can think of three likely questions associated with this plan:

✔ **What's the pool's perimeter?**

To find the perimeter, you add up the lengths of all the sides. You have two long sides of 10 m each, and four short sides of 5 m each. The long sides make a total of 20 m and the short sides make a total of 20 m, so the perimeter is 40 m.

✔ **What's the pool's surface area?**

To find the area, you split the shape into either a rectangle and a square or three squares. If you split it into three squares, they all have sides of length 5 m and therefore areas of $5 \times 5 = 25$ m^2, making a total of 75 m^2.

If you split it into a square and a rectangle, the rectangle has sides of length 5 m and 10 m, making an area of $5 \times 10 = 50$ m^2; the square has sides of 5 m and therefore an area of $5 \times 5 = 25$ m^2. Altogether, that makes 75 m^2, the same as the other way.

✔ **If the pool is filled to a depth of 1.2 m, what volume of water is needed?**

To find the volume, you times the area by the depth, so you do $75 \times 1.2 = 90$. You need 90 m^3 of water.

Chapter 15

Shaping Up

The difference between this chapter and Chapter 14 is that this one deals with moving shapes around while Chapter 14 deals with measuring shapes. I do include a bit of measuring in this chapter – and I suggest you have a protractor ready to measure the angles – but most of this chapter is about shapes and what you can do with them.

I remind you of some stuff you probably already know about shapes and what you can tell from just looking at them. I also take you through some of the words you need to be familiar with to deal confidently with shape questions.

Then I cover the joys of measuring angles (for example, to see how sharp a corner is) and get you messing about with fridge magnets (not included) to see how you can transform shapes. This leads on to the idea of symmetry – how you can transform a shape but end up with something that looks just the same. I also cover tessellation – making nice tile patterns with no gaps or overlaps – but to do this you need to understand how angles fit together.

I also run through the ideas of nets (unfolding a three-dimensional shape into a two-dimensional shape) and plans and elevations (what shapes look like from different angles – the top, the front and the side). Nets, plans and elevations aren't likely to come up in any numeracy exam, but your teacher might ask you to investigate them if you take classes.

You're Already in Good Shape

I reckon you're already familiar with all of the shapes you need to know about to meet the numeracy curriculum's requirements.

If you recognise squares, rectangles, triangles and circles, you're off to a good start. Add in cubes, cuboids (box shapes), spheres (ball shapes), cylinders and pyramids, and I think you're good on the 'recognising shapes' front. The shape of an object doesn't depend on its size or *orientation* – if you twist a square around, the shape is still a square, even if it looks like a diamond. Regardless of how big a shape is or which way around you draw it, a shape's properties and name stay the same.

The numeracy curriculum needs you to know what the terms 'between', 'inside' and 'near to' mean, but I won't insult your intelligence by explaining those.

You also need to know the difference between two-dimensional and three-dimensional shapes. The 'D' in a 3D movie stands for 'dimensions'. The dimensions make the movie seem as if everything isn't just flat on a screen but has depth. Similarly, in maths, two-dimensional objects are flat (you can draw them on paper) and three-dimensional shapes come out of the page.

Sussing out shapes you know

You need to recognise the following *two-dimensional* – or flat – shapes, which I also show in Figure 15-1:

- **Square:** A shape with four equal-length straight sides arranged at right-angles to each other.

- **Rectangle:** A shape with four straight sides at right-angles to each other. The sides aren't necessarily all the same length, but sides opposite each other are always the same length.

- **Triangle:** A shape with three straight sides. As you progress with maths, most of the geometry you do is based on triangles.

- **Circle:** This is the only curved shape you really need to know about. The technical definition is 'a shape with all of the points a fixed distance from the centre', but you'll recognise a circle when you see one.

Figure 15-1:
A square, a
rectangle, a
triangle and
a circle.

Three-dimensional shapes are shapes that don't lie flat. The following four three-dimensional shapes, which I show in Figure 15-2, are closely linked to the four major two-dimensional shapes that I describe above:

- ✔ **Cube:** The shape of a normal die. Each of a cube's sides is a square, and all of the edges are the same length.

- ✔ **Cuboid:** The shape of a shoebox. Each of its sides is a rectangle.

- ✔ **Pyramid:** The shape of . . . well, guess! It has a square on the bottom and four identical triangles around the side that come together and meet at the top.

- ✔ **Sphere:** The shape of a ball.

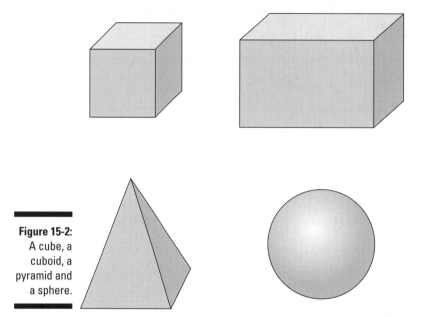

Figure 15-2:
A cube, a cuboid, a pyramid and a sphere.

Expanding your shapely vocabulary

In this chapter, I introduce some concepts that you may or may not have come across before. You may want to copy some of the definitions in this section into your notebook so you can work with them again later. Here are some of the words you need to know for the numeracy curriculum:

- **Angle:** The sharpness of a corner, measured in degrees. If you turn all the way around to face the way you started, you move through 360 degrees.

- **Reflection:** Turning a shape over.

- **Regular:** A regular shape is one where all the sides are the same length and all the corners have the same angle. A square is a regular shape, but you can have regular shapes with any number of sides, as long as it's three or more.

- **Right-angle:** A quarter-turn, or 90 degrees.

- **Rotation:** Twisting a shape around its centre.

- **Symmetry:** A *symmetrical* shape is one you can either fold precisely in half, or twist around its centre so you can't tell which way up it is.

- **Tessellation:** Putting shapes together to make a pattern, with no spaces between the shapes.

What's Your Angle?

Angle is another word for corner. For example, you may hear of footballers scoring from tight angles, meaning they've turned the ball round a sharp corner. The word 'angle' also shows up in other words, such as 'triangle' – which simply means 'three corners'.

Angles are measured in degrees. For example, if a car spins 360 degrees, it spins all the way round, while the latitude of London is 53 degrees north of the equator. Confusingly, angle degrees are completely different from temperature degrees – the context usually makes clear which type of degrees you need to work with (except possibly when talking about pointy icicles!). Both types of degree are denoted by a little circle above and after the number, for example: 90°.

Angles are interesting for many reasons, but one of the key points is that their properties don't really depend on how big the lines leading to the corner are. For example, the angle on a bookend is the same whether it's a bookend for tiny books or huge books – in either case, the angle is 90 degrees.

Defining angles

When someone says an angle is a certain number of degrees, they're trying to tell you how sharp the corner is. A small angle means the corner is very sharp, while an angle of 180 degrees isn't much of an angle at all but instead is a straight line.

One degree is defined as 'one three-hundred-and-sixtieth of a circle', which isn't tremendously helpful (but then again, you may not find the technical definitions of metres and kilograms helpful either). I prefer to think of 90 degrees as a right-angle – a sideways turn – and work from there.

Special angles

You need to know about the following special angles (which I show in Figure 15-3):

- ✔ **360 degrees** is a complete circle. If you turn around 360 degrees, you get back to where you started.

- ✔ **180 degrees** is half a circle. If you turn 180 degrees, you end up facing backwards.

- ✔ **90 degrees** is a quarter-circle, or a right-angle. If you turn 90 degrees, you end up facing to the left or the right of where you started.

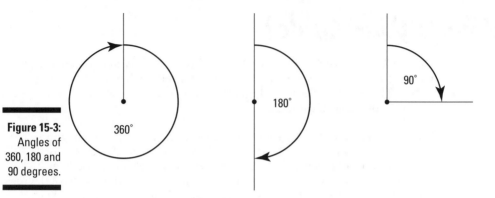

Figure 15-3:
Angles of
360, 180 and
90 degrees.

Other types of angle

You need to know the following angle-related words (I show examples in Figure 15-4):

- An *acute* angle is an angle smaller than a right-angle – so less than 90 degrees. I like to think of 'acute little puppy' to remind me that it's a little angle.

- A *reflex* angle is an angle bigger than 180 degrees. You know when your doctor hits the outside of your knee with a hammer to test your reflexes? I think of that to remember what a reflex angle is: the outside of your knee is always more than 180 degrees.

- An *obtuse* angle is in between – so more than 90 degrees but less than 180. I don't have a good memory aid for obtuse angles – so let me know if you can think of one.

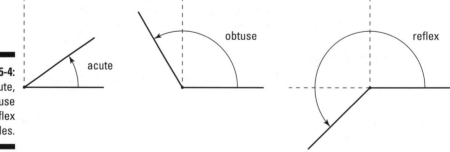

Figure 15-4:
Acute,
obtuse
and reflex
angles.

Measuring angles

You measure angles with a protractor, one of those semi-circular things you had in your pencil case at school but maybe only ever used to see if it worked like a Frisbee. (No, it doesn't. Or at least it didn't for me – it just got me into trouble for chipping my protractor and the wall.)

Here's how you measure an angle with a protractor (I show the steps in Figure 15-5 to help you visualise the set-up):

1. **Put the cross-hair in the middle of your protractor over the corner where you want to measure the angle.**

2. **Turn the protractor so that one of the lines going into the angle is on the 'zero' line across the bottom of the protractor, to the left of the cross-hair.**

3. **Follow the other line to the edge of the protractor and read the number off the scale – that's the angle you're looking for.**

Figure 15-5:
Measuring an angle with a protractor. The angle in this figure measures 27 degrees.

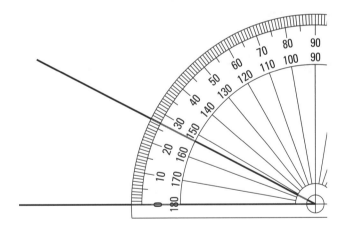

 Protractors are fiddly because they have two different scales running along the same edge and you can easily mix up which one is which. I recommend always using the outside scale – so make sure the zero you use is the zero on the outside track of the protractor.

 After you measure an angle, look at the angle again and ask, 'Does my answer make sense?' If you have an angle that's obviously more than 90 degrees (an obtuse angle) but your protractor says the angle measures 15 degrees, you know something has gone wrong.

Playing with Symmetry

Symmetry is a description of how you can change a shape but leave it the same. For instance, you can take a draughts piece and flip it over, and the piece looks just the same – this is an example of reflective symmetry. If you take a rectangle and spin it through 180 degrees, the shape looks like the same rectangle – this is an example of rotational symmetry.

If you've ever tried to solve a double-sided jigsaw, you may know that there are three things you can reasonably do to a jigsaw piece (if you don't count thumping the piece in a frustrated attempt to get it into a space). You can:

✔ Pick up the piece and move it around (the technical term for this is *translation* – you *translate* a shape when you move it).

✔ Turn it over (this is *reflection* – you *reflect* a shape when you turn it over).

✔ Turn it around (this is *rotation* – you *rotate* a shape when you twist it).

Reflection and rotation give you the types of symmetry you need to understand for the numeracy curriculum.

On reflection: Turning shapes over

A shape with *reflective symmetry* is one you can turn over without changing the shape. For instance, a rectangle has reflective symmetry: if you pick up a blank piece of paper and turn it over, you can put it back where it was before without anyone being any the wiser. (In fact, there are two ways you could do that: turn it over along the long edge or along the short edge.)

Squares, circles and some triangles also have reflective symmetry. The symmetry is described as reflective because one half of the shape mirrors the other. If you've ever played the mirror game where you stand facing the edge of a mirror so that when you lift your leg up it looks like you're standing in mid-air, you've used reflective symmetry.

A shape has reflective symmetry if you can draw a line on it somewhere so that one side of the line looks exactly like the other side of the line. Figure 15-6 shows some examples of reflective symmetry.

Figure 15-6: Reflective symmetry. Top left: a square has four lines of symmetry. Top right: a rectangle has two lines of symmetry. A circle has an infinite number of lines of symmetry (any line through the middle works). An equilateral triangle (with all sides the same length) has three lines of symmetry.

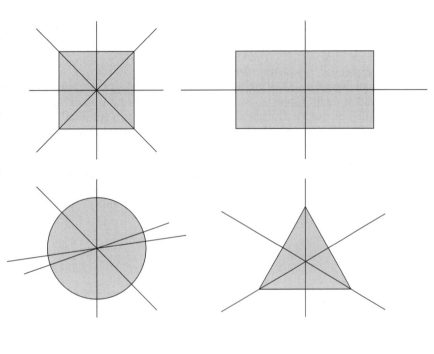

Doing the twist: Rotating shapes

If you do a normal jigsaw puzzle, the only way it makes sense to change a piece once it's the right way up is to turn it around so the bumps and spaces face another way, as I show in Figure 15-7.

Figure 15-7: Rotating a jigsaw puzzle piece: four ways around.

Rotating means twisting the shape around its middle so the shape faces a different direction. The piece in Figure 15-7 looks different every time you turn it, but some pieces look the same however you turn them around (ignoring the picture) – they either have all four of the bumps on the edges pointing inwards or all four pointing outwards. These pieces *have rotational symmetry*, meaning you can twist them around their centre and not know which way was originally up.

In general, the order of rotational symmetry of a shape is how many ways you can twist it round to get the same shape before you get back to the beginning, as I show in Figure 15-8 – or how many different ways you can hold the paper and see exactly the same shape. A square has rotational symmetry of order four, because you can hold the paper the normal way up, upside down, facing to the left or facing to the right, and the square will look just the same. A rectangle has rotational symmetry of order two, because it looks the same the normal way up and upside-down – but if you turn the paper sideways, it looks different. A triangle, depending on the type, may have order three, order two or no rotational symmetry. And a circle . . . that's a trick question: a circle has infinite rotational symmetry.

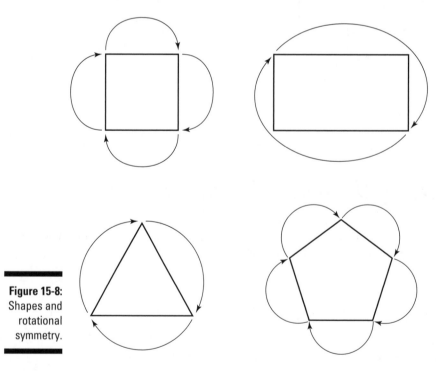

Figure 15-8:
Shapes and
rotational
symmetry.

Pretty patterns: Tessellation

Tessellation is a fancy word for fitting shapes together so that there are no gaps between the shapes and none of the shapes overlap – as if you're solving a jigsaw puzzle, tiling a wall or paving a path. It may seem like there's not very much maths involved in tessellation, but in fact it's all about the angles.

Sad fact for the day: once upon a time, I was number 17 in the world at the Linux version of the game Tetris. I find it difficult to think of another achievement that's simultaneously so impressive and so pointless.

The idea of Tetris is to manoeuvre different-shaped blocks falling from above into a rectangular space – any time you complete a line all the way across without any gaps, the line disappears and the blocks above it shift down.

Tetris is all about tessellation: fitting shapes together so there aren't any gaps. Some other places you see tessellation are in the work of Dutch artist M.C. Escher and in a great deal of Islamic art – for instance at the Alhambra Palace in Spain.

You don't need to think of tessellation in quite the same depth as these artists – although it can be fun to do so. For the numeracy curriculum, you only need to worry about fairly regular shapes. By the way, tessellation is part of the syllabus, but setting multiple-choice questions on tessellation is very hard, so it doesn't often come up in the exam.

Tessellation has one important rule: wherever lines meet, the angles have to add up to 360 degrees.

Tetris works because the corners on all of the shapes are 90-degree angles, and when four of the shapes meet you end up with no spaces, as you can see in Figure 15-9. Not only 90-degree angles tessellate, though. To give just a few examples, you can also tile equilateral triangles (with 60-degree corners) and hexagons (six sides and 120-degree corners), also shown in Figure 15-9.

The only kind of multiple-choice question I can think of that would involve tessellation involves 'filling in the gap' – the examiner gives you two or three shapes that meet at a corner and you need to find the angle on the remaining shape.

This is a pretty simple process if you remember the important rule that I mention above:

1. **Write down the size of each angle touching the corner you're interested in.**

2. **Add up all the angles from Step 1.**

3. **You need to make 360 degrees in the corner. Work out: 360 take away the angle you worked out in Step 2. The answer is the size of the angle you need to put in the corner.**

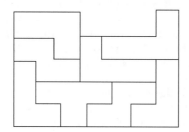

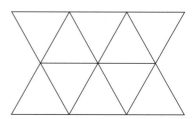

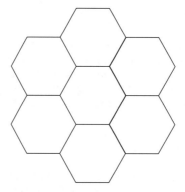

Figure 15-9:
Tessellation.
Tetris
blocks,
triangles
and
hexagons
tessellating.

Understanding Nets, Plans and Elevations, Oh My!

I've been told at least once that I'm the least artistic person on the planet, which rankles a bit: I *could* learn how to draw, I just haven't yet.

Maybe you have all of my missing artistic skills – I hope so, because the numeracy curriculum suggests you should be able to investigate the following drawing-related ideas:

- **Net:** A picture of what a 3D shape looks like if you unfold it.
- **Plan:** A top-down view of a 3D shape.
- **Elevation:** A side-on view of a 3D shape from any side.

At this stage, you don't actually need to draw these things, although if you study maths further you may do. In this section I just give you a few ideas about what you can do with nets, plans and elevations.

Folding under pressure: Nets

A *net* is a shape you can fold up, origami-style, to make a three-dimensional shape without any gaps. Nets are really hard to visualise, but you only need to know a few shapes for the numeracy curriculum. Figure 15-10 shows nets of a few common shapes. Notice the difference between the net of a cube (six squares) and the net of a pyramid (a square with four triangles next to it). The net of a cone looks a bit like Pacman.

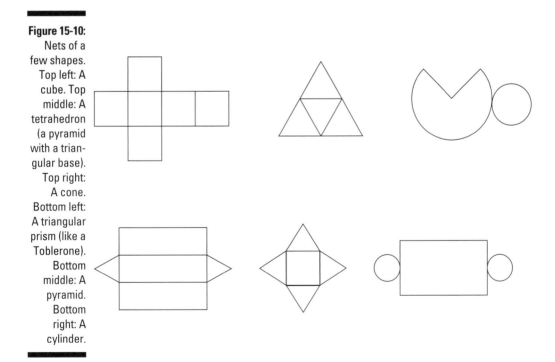

Figure 15-10: Nets of a few shapes. Top left: A cube. Top middle: A tetrahedron (a pyramid with a triangular base). Top right: A cone. Bottom left: A triangular prism (like a Toblerone). Bottom middle: A pyramid. Bottom right: A cylinder.

Looking at every angle: Plans and elevations

Architects like to show you what the thing they're building looks like from the front, back and sides, and sometimes the top, in case you need to pick it out on Google Earth or locate it from an aircraft.

Being able to sketch what an object looks like from the front and sides – and conversely, to 'see' from those views what the object looks like in 3D – is unlikely to help you in an exam, but you may need to do it in an investigation if you take maths classes.

Understanding plans and elevations is also a useful step towards understanding 3D shapes for further study. Being able to visualise what's going on, and having a good sense of up, back and sideways can make the shape part of GCSE maths more accessible and engaging.

To draw a *plan* – a top view – you imagine what you'd see if you looked down on your shape. What shape is the top? Would you see any lines? How big is the top? Then you sketch what you see. For the *front elevation*, you do the same thing, looking from the front. And for a *side elevation*, you draw what the side looks like.

When doing elevations and plans, *don't* take perspective into account. If you look at the front of a house where the roof slopes directly back, your eyes see the roof sloping inwards but the elevation shows the roof going straight up. Have a look at Figure 15-11, with a couple of examples of elevations and plans, to see what I mean.

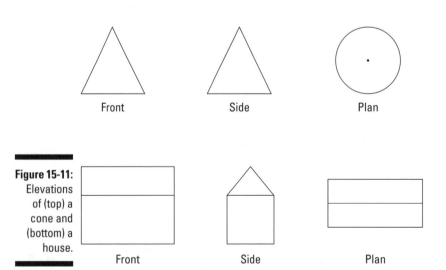

Front   Side   Plan

Figure 15-11: Elevations of (top) a cone and (bottom) a house.

Front   Side   Plan

Part IV
Statistically Speaking

'I was very disappointed with the latest job applicants & their knowledge of Basic Maths — 27% were only average & the other 76% were hopeless.'

In this part . . .

Graphs and tables are a fantastic way to get information across efficiently, letting you see the big picture as well as the detail. In this part, I show you how to read these beasts and create your own, both by hand and using a computer.

I also run you through the different types of statistic you're supposed to know about – the three averages (mean, mode and median) and the range, before regaling you with tales of probability. Exciting, eh?

Chapter 16

Data Mining (No Hard Hat Required)

*B*eing able to read and understand tables and graphs is one of the most valuable real-life skills you can conceivably pick up from reading this book. In this chapter I introduce you to most common types of graph. I help you read and understand bar charts, pie charts and line graphs, and show you which ones work best in different situations.

I devote a fair chunk of this chapter to tables, which you may find easier than graphs – but beware, because even simple tables have pitfalls to watch out for. I also cover some of the harder questions that may come up in your maths exam relating to graphs and tables, such as finding differences, drawing conclusions, reporting the bigger picture, and finding mistakes – something that even professional mistake-finders like me find difficult.

A Spotter's Guide to Graphs and Charts

You use graphs and charts to communicate large amounts of data in a compact and organised way, normally using visual effects to show the big picture. For example, it's much easier to see that sales are growing from a graph that gradually rises than it is from a list of figures. In the same way, a pie chart could quickly show you where you're spending most of your money in a more obvious way than your bank account does.

Graphs and tables come in four basic varieties:

▶ A *number table* does exactly what it says on the tin. It's a table . . . with numbers in. A good example is the list of temperatures in cities around the world you sometimes see in the newspaper.

✔ A *bar chart* looks a little like a picture of a skyline, consisting of different heights of 'tower' lined up side–by-side. The heights of the towers represent the relative sizes of the categories they represent. Some graphs show the towers on their sides, but you can deal with those by tilting your head or turning the paper through 90°. (Incidentally, if you don't know what 90° means, head to Chapter 15, where I give the low-down on angles.) You could use a bar chart to show the typical fuel consumption of different types of car.

✔ A *pie chart* looks – you've guessed it – a bit like a pie. The chart is a circle with various-sized slices 'cut out' from the middle to the edge. The size of the slices shows the relative size of the categories. You often see a pie chart showing the results of an opinion poll.

✔ A *line graph* shows how a value changes, usually over time. Most line graphs look like a jagged line going across the page. How high the line is above a time marked on the axis tells you how high the value is. A dieter may use a line graph to track how their weight fluctuates as time goes by. A business may use a line graph to track its profits.

Nailing number tables

When my parents were at school, calculators weren't common – and those that existed were the size of a car. (Your mobile phone probably has more computing power than existed in the entire world 40 years ago.) Instead of working things out on a calculator, my parents looked up answers in a huge book full of numbers. They had to find the right page, and then find the right row and column, and copy the number down carefully – and assume that the printer hadn't made a mistake anywhere. Oh, and then do the sum.

Your task, thankfully, is less of an ordeal. For the numeracy curriculum, you only use tables with a few rows and columns. You still have to find the right cell in the table and do the sum, but much less can go wrong than could have done for maths students of my parents' generation.

A number table normally has a *label row* at the top and a *label column* on the left, which give you information about each *cell* (square) in the table.

The Table of Joy, which I introduce in Chapter 8 and use in most chapters in this book, is a number table. It has a label row at the top and a label row on the left, and numbers in all the other cells. The labels tell you what each cell represents.

You use a number table if you want to look up a specific answer. You often use a table if you have numbers that depend on two other things. For instance, the price of a hotel room may depend on both the luxury of the room and what day of the week you want to stay – so you may represent hotel-room prices in a table.

Another example of a number table is a mileage chart showing the distance between pairs of cities. Each of the numbers in the table depends on two things – the two cities.

In comparison, a graph tends to show an overall picture of the data. For example, if you want to show the monthly average temperatures in several cities around the world, you may use a table ('I'm going to Toronto in April and want to look up what temperature to expect') but also a bar graph ('I want to see which of these cities is generally the warmest'). See the next section on 'Bringing in the bar charts' for more on graphs.

Bringing in the bar charts

Bar charts are made up of a series of rectangles – or *bars* – with each bar representing a different group. Each bar has the same width, but the heights vary to show the 'value' of each bar's group – for example, how many people are in the group, or how much money comes from the group, or how many goals the team scored.

Knowing when to use a bar chart

You use a bar chart to compare the values of several numbers at once. The numbers could be measurements, amounts of money, numbers of people or things, all in different groups or categories. It lets you see at a glance which group is the most or least important.

You can also figure out the actual value of each bar by drawing a horizontal line across to the scale and reading off the number.

Bar charts are a little more honest than pie charts, which I describe in the section 'Poking about in pie charts'. The human brain is much better at assessing distance than it is area and angle, so we perceive bar charts more accurately than we do pie charts. Calculating the actual value of each group on a bar chart is also easier than on a pie chart.

You use bar charts and pie charts in similar situations – when you have several, separate values attached to separate categories. One situation where you definitely use a bar chart rather than a pie chart is when looking at the numbers as part of a whole doesn't make sense.

Bar charts are better than line graphs when your data are in several distinct groups rather than over something measurable like time.

Good examples of when to use a bar graph are to compare the house prices in several different regions of the UK, to show the personal best race times of several sprinters, or to compare the heights of different species of tree.

Single-bar charts

The simplest kind of bar chart is called, a little misleadingly, the single-bar chart. I say 'misleadingly' because any bar chart worth its salt has at least two bars in it, otherwise it's not really comparing anything and you may as well just have written the number down. The word 'single' means that each category has only one number associated with it, so you get a single bar for each category.

Single-bar charts are by far the easiest type of bar chart to read and understand. To find out what a bar represents, here's what you do:

1. **Get a ruler (or anything with a straight edge) and lay it flat across the top of the bar, going sideways across the graph.**

2. **Make a small mark where the ruler crosses the vertical axis (the vertical numbered line).**

3. **If the mark lies on a value given on the scale, that's the value of the category represented by the bar.**

4. **If the mark lies between two values, make an estimate of the number.**

   Think about whether the number's halfway between the neighbouring values, or a little more or less.

I show an example of a single-bar chart in Figure 16-1.

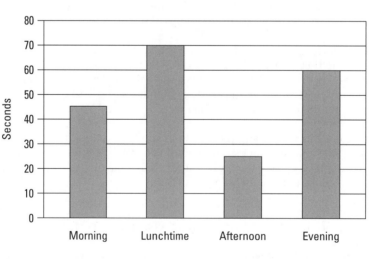

Waiting times

Figure 16-1:
A single-bar chart showing average call response times at a made-up company.

Multiple-bar charts

More complicated – and sadly, more common in exams – are multiple-bar charts. You use multiple-bar charts to compare two different values across categories. For example, to investigate exam pass rates of several schools, you may want to compare the results of boys and girls. You then have two distinct bars in each category, coded with shading, as in Figure 16-2.

Another example is using a multiple-bar chart to compare average summer and winter temperatures in several cities.

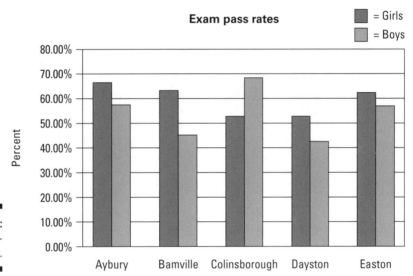

Figure 16-2: A multiple-bar chart.

Reading a multiple-bar chart is very similar to reading a single-bar chart. The only difference is that you need to make sure you look at the correct bar. Before you start, check the *key* – the little box that tells you which colour or type of shading corresponds to which subcategory – and then look at the bar in the correct category with the right colour or shading.

Stacked-bar charts

I nearly forgot to mention stacked-bar charts, because I dislike them so much. The purpose of these monstrosities is to show the changes in both the totals and the composition of a value – for instance, not only how your company's income has changed but also how much of the income has come from each source. I give an example of a stacked-bar chart in Figure 16-3.

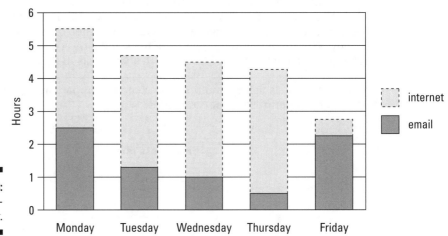

Time spent on email and internet

Figure 16-3:
A stacked-
bar chart.

Some of the information in a stacked-bar chart is pretty easy to read – you can figure out the total value and the value of the lowest bar using the same methods you use for normal bar charts. The values in the middle and at the top are a bit trickier. Here's what you do:

1. **Work out the value of the bottom of the bar you're interested in by measuring across to the vertical axis, as you do with a regular bar chart.**

2. **Work out the value of the top of the bar you're interested in, using the same technique.**

3. **Take the answer in Step 1 away from the answer in Step 2.**

   The result is the value of the bar.

I don't like stacked-bar charts because they're a mess. Comparing two bars next to each other when their base has moved is really hard, so be careful. I'm not the world's biggest fan of multiple-bar charts either, but I think they're much clearer than stacked-bar charts.

Poking about in pie charts

Pie charts use angles to show the relative sizes of various categories. Pie charts are circular and are cut into 'slices': the bigger the slice of pie, the bigger the group it represents. I give an example of a pie chart in Figure 16-4, showing you roughly how I spend my time during a typical day.

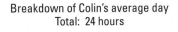

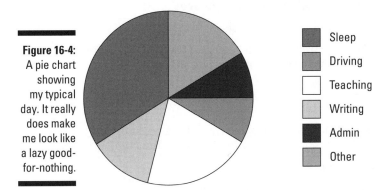

Breakdown of Colin's average day
Total: 24 hours

Sleep

Driving

Teaching

Writing

Admin

Other

Figure 16-4:
A pie chart showing my typical day. It really does make me look like a lazy good-for-nothing.

Deciding when to use a pie chart

You use a pie chart rather than a bar chart when you're not very interested in the actual numbers you want to represent but want to see how big the groups are compared with each other. A good example is if you want to give a presentation about the age groups of your company's customers but don't want the audience to know precisely how many customers you have. You frequently see pie charts on election-night reports on TV to show the distribution of votes. In a very close election race, the slices representing the two front-runners are almost the same size.

Handling angles, percentages and numbers

If you've read many other chapters in this book, you probably won't be surprised to find that you can work out the values associated with pie charts using the Table of Joy – and if you don't know what I mean, head to Chapter 8, where I explain the joy of the Table of Joy in great detail. A whole circle contains 360 degrees (as I describe in Chapter 15). In a pie chart, those 360 degrees correspond to the total of the values represented in the chart. When you work with a pie chart, you may need to figure out one of the following three things:

✔ The size of the angle in a slice.

✔ The value of a slice.

✔ The total of the values in all the slices.

To find one of these things, you need to know the other two. Here's how to use the Table of Joy to work with a pie chart:

1. **Draw out a noughts-and-crosses grid.**

   Leave yourself plenty of room in the grid for labels.

2. **Label the top row with 'value' and 'degrees'.**

   Label the sides with 'slice' and 'circle'.

3. **Write 360 in the 'circle/degrees' cell and the two other pieces of information you have in the appropriate places.**

   Put a question mark in the remaining cell.

4. **Write down the Table of Joy sum.**

   The sum is the number in the same row as the question mark times by the number in the same column, all divided by the number opposite.

5. **Work out the sum.**

   The answer is the value you're looking for.

To convert an angle into a percentage (or vice versa) you use a similar process. The whole circle – 360 degrees – corresponds to the whole of the data – 100 per cent. Use the same steps as above, but change the 'value' column to 'per cent', and in the 'circle/per cent' cell write 100, just like the example in Figure 16-5.

Figure 16-5: Converting percentages to pie-chart angles with the Table of Joy.

	Percent	Angle
Circle	100	360
Slice	30	?

$$\frac{30 \times 360}{100} = 108$$

Looking at line graphs

Line graphs are probably my favourite kind of graph.

The idea of a line graph is to show how a value changes in response to another value – often, but not always, time.

In Figure 16-6 I give an example of a line graph showing the world population. Notice the zigzag on the vertical line – or *axis*, showing that the numbers don't start at zero. At first glance, it looks like the population has doubled, but in fact it's 'only' increased by about 20 per cent.

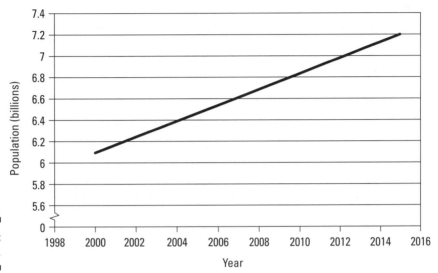

**Estimated world population
2000 - 2015**

Figure 16-6:
Line graph.

Careful with those axes!

Graphs are supposed to make data easier to visualise and understand – and, to a large extent, they do. But obtaining precise data from a graph is still sometimes tricky.

Here's the process for reading a line graph when you know a value on the horizontal axis:

1. **Find the relevant value on the horizontal axis.**

   The value may be marked. If it's not, try to estimate where the value should be: find the two values it sits between and decide which one it's closer to. Make a little mark on the axis there.

2. **Using a ruler, lightly draw a straight line in pencil directly up from the mark until it reaches the graph.**

3. **Now turn the ruler a quarter-turn and draw (still lightly in pencil) across from where your vertical line meets the graph, until you reach the vertical axis.**

4. **Where this line meets the vertical axis is your answer.**

   If the line is on a marked value, fantastic – that's your answer. If your line doesn't quite sit on a marked value, do a bit of inspired guesswork. Which values is your line between? Is it about halfway between them, or closer to one than the other?

TIP

Make your marks lightly in pencil because you may need to look at the graph again to find more values.

If, instead, you have a value for the vertical axis, you simply work the other way around: you find the value on the vertical axis, draw across to the graph, and then draw down to the horizontal axis, where you read off the value you need.

More than one line

Sometimes you have two or more lines to work with. You read the values off the graph in the same way as for a graph with one line. The only difficulty is making sure you use the correct line.

Before you start, look for the *key* or *legend* – a little area containing information about the graph. The key shows what the different colours or styles of line represent. Your mission is to pick the line that best suits what you're looking for.

I show a multiple-line graph with a key in Figure 16-7.

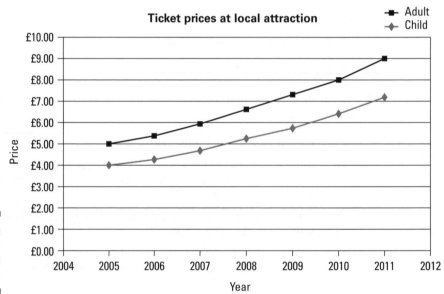

Figure 16-7: A multiple-line graph with a key.

Ch-ch-ch-changes

A line graph shows the change from one time period to the next really well. If the graph slopes up and to the right, the value is increasing. If the graph slopes down and to the right, the value is falling.

Try remembering that *upright* citizens are held in *high* regard, while *downright* dirty dogs are held in *low* esteem. Daft – but it works for me.

Here's how to work out by how much a value has risen or fallen between two time periods:

1. **Read the value for the first time period on the horizontal axis.**

   Use the same method as I explain in the section 'Careful with those axes!'

2. **Read the value for the second time period in the same way.**

3. **Take the smaller number away from the bigger number.**

4. **If the value has fallen, put a minus sign in front of the number.**

   The answer is how much your value has risen or fallen.

In a multiple-choice exam, read graph questions very carefully. Examiners are often sneaky and put in positive and negative versions of the same number. Make sure you get the sign right! For more on negative numbers, check out Chapters 3 and 13.

Reading Graphs, Tables and Charts

To interpret a good, properly drawn graph, you don't need special skills. What's going on should be obvious from the picture.

Unfortunately, most graphs you see aren't drawn properly. They're too busy, or not labelled clearly, or generally a mess. I give you some tips on how to make sure your graphs don't fall into this trap in Chapter 17. Here, in this chapter, I focus on how to gather information from graphs, tables and charts – which may not be as well laid out as the ones you produce.

Picking the right data from a table

Many tables suffer from information overload. A big, complicated table, no matter how carefully produced and well set out, is intimidating and difficult to handle.

The more slowly and carefully you deal with tables, the better. Write down what you want to find out – even if that means copying out an exam question. Writing like this forces you to slow down and think about what you need to know.

Finding the right row and the right column

When you know what you want to find, you can start looking for it. Sift through the labels at the top and at the side of the table to see which labels match what you're looking for. Then read down and across – the number you need is in that row and column.

I put a ruler along the right-hand side of the column I'm interested in, so when I read across a row I know to stop when I reach the ruler.

Matching criteria

In an exam, a common type of table question involves finding values that match a criterion – where the number is greater than or less than a given value. For instance, after reading an exam question, you may know a business's sales for each month of the year and that the business needs to sell at least 100 widgets each month to make a profit. To find how many profitable months the business had over the year, you simply count how many of the entries in the 'widget sales' column are over 100.

Doing further sums

Sometimes you need to think a little bit laterally. For instance, an exam question may ask for the difference between two values, or may ask for their total. In this case, you need to find both of the values and then do the sum. Exam questions often ask about statistical properties of tables – the range, mode, mean or median of data. I cover all of these in Chapter 18, so feel free to skip ahead if you're curious.

Keeping up with keys and axes

Being able to read graphs depends critically on being able to understand the keys and the axes.

The *key* or *legend* to a graph is a set of information next to the graph that tells you what represents what – for instance, which colour in the graph represents which category, or which type of line means which set of data. The key is the first place you should look when you answer an exam question on graphs, as the key tells you which part of the graph is important to you.

The *axes* are the horizontal and vertical lines at the bottom and side of the graph. I show you how to use axes in the section 'Bringing in the bar charts'. Axes are very important: reading the numbers off an axis is the single most important step in answering graph questions.

Understanding graphs

Well-drawn graphs can be fantastically useful for communicating information on two levels:

- ✔ **Big picture:** You can see just by looking at a line graph whether a value is increasing, decreasing, staying the same or wobbling all over the place, much more easily than you can with a table. A bar chart or pie chart shows you which categories are the most important at a glance.

- ✔ **Detail:** With a little work, you can extract the numerical value of each category from a graph. This opens up a whole range of ways to interpret and describe the data in the graph.

Drilling Deeper into Graphs

If all you ever had to do with graphs was read off numbers, I think graphs would be one of the least complicated topics in the numeracy curriculum. Unfortunately, you frequently have to do a little bit more with graphs in maths exams, such as finding the difference between two values on a graph, counting how many times one value is bigger than another value, summarising graphs, or finding trends such as generally increasing or generally decreasing.

Adding up totals and finding differences

This kind of exam question often lurks near a graph, waiting to trip up an unsuspecting passer-by. The sums you need to do aren't particularly difficult – the hard thing is picking out which points on the graph you need to work with.

Difference questions ask things like 'How much more . . .' or 'How much less . . .' or, if you're lucky, 'What's the difference between . . .' The trick is to find the relevant points on the graph, figure out the values of those points, and then take the smaller numbers away from the bigger numbers.

In Figure 16-8 I show you how to use the graph from Figure 16-7 to answer the question 'How much more did an adult ticket cost in 2008 than a child ticket?' An adult ticket cost about £6.60 and a child ticket £5.30, so the difference is £1.30.

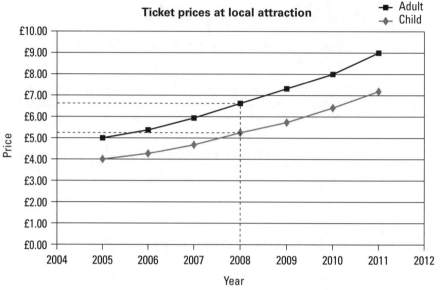

Ticket prices at local attraction

Figure 16-8:
An example of a difference sum with graphs. Read off the values you're interested in and take the smaller values away from the bigger values.

A special case of a difference question is 'By how much did a value increase over the whole time period?' For this type of question, look at the numbers at the start and the end of the time period and take the smaller number away from the bigger number.

Total-type sums are marked by words such as (surprise!) 'total', 'sum' and 'altogether'. The only difference between total sums and difference sums is that you add instead of take away. You find all of the points that fit the description given in the question, and add up their values.

In Figure 16-8 I show you how to figure out the total cost of taking a family of two adults and three children to an attraction in 2008. Each adult cost £6.60 – so two adults cost £13.20. Each child cost £5.30, so three children cost £15.90. Altogether, that's £29.10.

Summarising graphs

Give a graph to a dozen statisticians and they'll probably give you a dozen different answers about what they think is the best way to describe the graph, just like a dozen art critics give a dozen different interpretations of a masterpiece.

I don't mean there's no right and wrong when looking at graphs. Rather, there are many things about graphs that are true (and also many things that are false). Summary questions in exams are awkward: usually all of the options in a multiple-choice summary question are pretty plausible and you can't just say 'That one's correct', even if you're an experienced graph-examiner.

In this section I offer some tips about how to approach this kind of question:

- ✔ If an option mentions a trend, decide whether the graph seems to be moving upwards or downwards over time.

- ✔ If a multiple-choice answer talks about something increasing or decreasing in every time period, whether that's months or days or years, check that *every* line segment on the graph slopes upwards and to the right (for increasing) or downwards and to the right (for decreasing) – if one is flat or going the other way, this is the wrong option.

- ✔ If you want to look at the amounts of increase and decrease, try to work out these amounts using the ideas in the section 'Adding up totals and finding differences'. A multiple-choice option may be 'this value halved' or 'this value doubled' – in which case, check whether the end value is half as much or twice as much as the start value.

- ✔ A graph showing an average probably tells you nothing about the range of the values. Unless you can see a separate line for the highest and lowest values, any multiple-choice option that talks about the range or the extreme values is almost certainly the wrong one.

Catching errors

One of the questions that comes up over and over again in numeracy exams is 'What's wrong with this graph?' Most people find this question annoying, but don't panic – the examiner can do only a limited number of wrong things to a graph. Try looking for the following errors:

- ✔ **Missed or mis-plotted data points:** If you have both a table and a graph, check all the relevant points from the table are also on the graph. Are the points in the right place on the graph? Do the sizes of the bars in the bar chart match up with the numbers in the table?

- ✔ **Incorrect labels:** Do the numbers and the categories match up? Is the title accurate? Are the sectors of a pie chart labelled, or is there a key?

- ✔ **Incorrect scales:** Do the numbers on the vertical axis start at zero?

- ✔ **Missing total:** An honest pie chart shows the total number, so try to work out the raw numbers.

Chapter 17

Top of the Charts

. . . .

In This Chapter

▶ Making tables

▶ Drawing bar charts

▶ Perfecting pie charts

▶ Learning to love line graphs

. . . .

*Y*ou may have heard of a famous book – famous in maths circles at least – called *How to Lie with Statistics*. The book's full of all manner of sneaky tricks to make your graphs look more convincing than they really are. Most people understand and trust pictures much more readily than they do numbers, and graphs are simply a visual summary of whatever data you're choosing to represent. Some of the tricks in *How to Lie* come up in numeracy exams. The only difference is that examiners call the tricks 'errors'.

In this chapter I show you how to set up a table of data, in preparation for making your own graphs as misleadingly or as honestly as you like. I can think of as many ways to make a table as there are people making tables, so I just show you a simple, reliable way in this chapter. Feel free to experiment with your tables and make them as awesome as you like.

Multiple-choice exams can't demand that you draw a graph, so the questions are normally limited to asking about which graph to choose and the details of the drawing process. If you do a written exam though, you may need to draw your own graphs.

I also help you pick the right graph to draw and show you how to draw it correctly. In this chapter I cover three kinds of graph – bar charts, pie charts and line graphs. You may want to read Chapter 16 first to make sure you understand the differences between the different graphs.

Turning the Tables

When you want to present some information based on two categories, using a table is almost certainly your best bet. You put one category label across the top of the table, and the other category label down the left-hand side. The number in any box of the table – or *cell* – then corresponds to both the label at the top and the label at the side. In Figure 17-1 I show you the various bits of a table.

For example, a bus timetable has two categories: the numbered or named bus route (which normally goes across the top) and the stop you're interested in (down the left-hand side). To find out when a particular bus reaches a particular stop, you work down the column for the bus until you reach the row for the stop. Or you can work the other way around: move along the row for the stop until you reach the column for the bus you want to catch.

For more examples of tables, have a look at the section on reading tables in Chapter 16. In this section, I concentrate on making your own tables on paper and on a computer.

Quantities being measured

	Area (km^2)	Volume (km^3)	Length (km)	Mean depth (m)
Lough Neagh	383	3.53	30	9
Lower Lough Erne	109.5	1.3	29	11.9
Loch Lomond	71	2.6	36	37
Loch Ness	56	7.45	39	132
Upper Loch Erne	34.5	0.35	19	2.3
Loch Awe	39	1.2	41	32
Loch Maree	28.6	1.09	20	38
Loch Morar	27	2.3	18.8	87
Loch Tay	26.4	1.6	23	60.6
Loch Shin	22.5	0.35	27.8	15.5

Things (lakes) being measured

Values for each lake/measurement

Figure 17-1: The anatomy of a table.

If you've worked through any examples in other chapters of this book, I bet you've written out at least one Table of Joy. For more on this delightful table, have a look at Chapter 8. You may also have seen the more in-depth times and addition tables in Chapters 3 and 4.

Making your own tables

You follow the same method for making pretty much any number table:

1. **Pick the two categories you want to work with.**

2. **Decide which category to put across the top and which to put down the side.**

   Usually which category goes where doesn't matter. I tend to put the category with more entries down the side, so the table fits on the page.

3. **Draw the table.**

   You need one more column than you have labels across the top, and one more row than you have labels down the side. This is so you have room to write the labels.

4. **Label the columns.**

   Leave the top-left square blank. Then, starting in the next square to the right, fill in the labels across the top.

5. **Label the rows.**

   Leave the top-left square blank. Then, starting in the next square below, fill in the labels down the side.

6. **Fill in the numbers in the correct squares.**

You probably realise that making a table is hardly rocket science. The main difficulty comes in dealing with the excruciating boredom of filling out row after row of numbers accurately. I've done data entry jobs in the past. They're no fun.

Making tables is always easier on a computer than on paper. If you can use a spreadsheet program such as Microsoft Excel, you'll find making tables a lot quicker and probably neater than working by hand.

Looking at a real-life table

In this section I help you create a real-life table using a bed-and-breakfast business as an example.

Imagine you run a bed and breakfast. You offer a standard room at £50 a night and a deluxe room at £70 a night. You also charge an extra £15 for Friday and Saturday nights. You could explain all those prices in words . . . or you could use a table.

You choose 'Days' as your category for the top and 'Room type' as your side category. Alternatively, you can have the days along the side and the rooms along the top. You have two types of day and two types of room, so you need three columns and three rows. Now draw the table: label the days 'Sunday–Thursday' and 'Friday and Saturday', and label the rooms 'Standard' and 'Deluxe'. Then fill in the numbers: from Sunday to Thursday a standard room is £50 and a deluxe room is £70; at the weekend, the standard is £65 and the deluxe £85.

Have a look at Figure 17-2, which shows what your final table should look like.

Figure 17-2:
A bed-and-breakfast table.

	Sunday to Thursday	Friday and Saturday
Standard room	£ 50.00	£ 65.00
Deluxe room	£ 70.00	£ 85.00

Tallying up

A *tally chart* is a way of recording information in categories very quickly without having to think too hard. Tally charts are perfect for taking surveys or categorising information when you're in a hurry.

Figure 17-3:
A completed tally chart showing the favourite seasons of 40 of my imaginary friends.

Season	Tally	Total
Winter	⦀⦀ ⫼	7
Spring	⦀⦀ ⦀⦀ ⏐	11
Summer	⦀⦀ ⦀⦀ ⏐⏐⏐	13
Autumn	⦀⦀ ⏐⏐⏐⏐	9

Here's what you do to make a tally chart like the one in Figure 17-3:

1. **Draw a table with three columns.**

   Make the left-hand column wide enough for labels and the middle one as wide as you can. Label the columns with the category name, 'Tally' and 'Total'.

2. **Make a row for each category you're counting and label it in the left-hand column.**

3. **Do your observations.**

   Go through your list or do your survey. Each time you see a person or thing in a particular category, put a little mark in the appropriate middle column. Be neat; keep your marks evenly spaced and in a straight line.

4. **Group your marks into fives.**

   The traditional way of doing this is to make every fifth mark in a category a diagonal line through the previous four, as in Figure 17-3. You can also wait until the end and just put a diagonal line through every group of five; it's entirely up to you.

5. **Count up your tally in each category.**

   In each row, count the number of groups of five and times that number by 5. Add on any loose marks at the end and write the number down in the final column.

Watching out for problems

The main thing to look out for when you make a table is boredom. Slipping up and missing out a number is very easy – but a very hard mistake to catch unless you continually check your table as you go along.

My best strategy for not missing numbers out is to check at the end of each row whether you're in the right place. If the labels at the top and the start of the row match up with what you're trying to communicate, you're probably okay. If the labels don't quite match the information in the cell, go back and look at that whole row again.

Another trouble-spot is missing out an entire row of a table by not looking carefully. Missing out information can be embarrassing, so always double check your table.

Grappling with Graphs

As far as the numeracy curriculum is concerned, you need to know about three kinds of graph – bar charts, pie charts and line graphs. I cover these graphs in detail in Chapter 16, so you may want to read that chapter before reading the rest of this section.

A *bar chart* looks a bit like a picture of a skyline – a jumble of skyscrapers of different heights. Each skyscraper – or bar – represents a thing or group of things. In the bar chart in Figure 17-4, each bar represents a type of music in my collection. The height of each bar shows you how many of each type of CD I have.

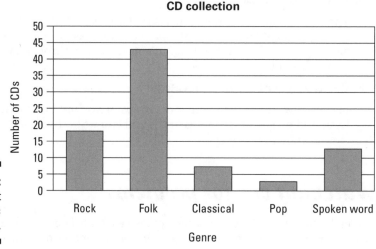

Figure 17-4:
A bar chart
of my music
collection.

A *pie chart*, rather surprisingly, looks like a pie – with slices (usually of different sizes) arranged in a circle. Each slice of the pie represents a category or group of things. The size of each slice (strictly, the angle at the centre) shows the relative size of the category. The pie chart in Figure 17-5 gives the same information about my CD collection as Figure 17-4 and shows the relative sizes of each genre of CDs.

CD collection

Total: 84 CDs

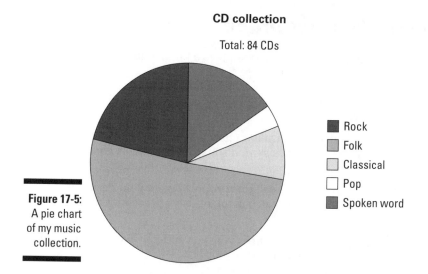

Rock
Folk
Classical
Pop
Spoken word

Figure 17-5:
A pie chart
of my music
collection.

A *line graph* shows you how some quantity varies when something else changes. If you watch a TV news programme, you may see a line graph showing how the stock exchange has behaved recently, or how crime statistics, house prices, unemployment rates and so on have changed over time. How high up the line is shows the value of the thing you want to measure at any given value of the thing on the bottom. In Figure 17-6, you can see a line graph showing the percentage turnout at UK general elections since 1945.

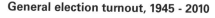

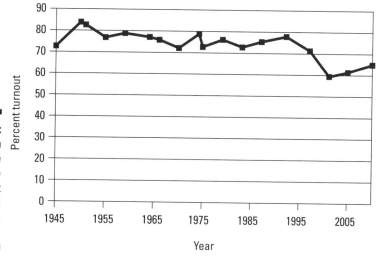

General election turnout, 1945 - 2010

Figure 17-6:
Line graph
showing the
percentage
turnout at
UK general
elections
since 1945.

Picking the right graph

Picking the best graph to show your results is a bit of a dark art. Here are my tips for choosing the best graph:

- Use a bar chart to show the actual size of your values and to compare the different values.

- Use a pictogram if you think a bar chart is a bit boring. A *pictogram* is just like a bar chart but uses symbols instead of bars, as shown in Figure 17-7.

- Use a pie chart to show the relative proportions of your data – in roughly the same situations as you might use a percentage to describe them.

- Use a line graph if your data vary over time and you want to show the changes between time points.

- Use a scatter plot if your data vary over two categories. For instance, you can use a scatter plot to show how people's shoe sizes are related to their heights. A *scatter plot* is simply a bundle of crosses plotted in the appropriate place on a graph – you don't need to be able to draw or read a scatter plot for your numeracy exam, but you do need to know what a scatter plot is. I show an example in Figure 17-8.

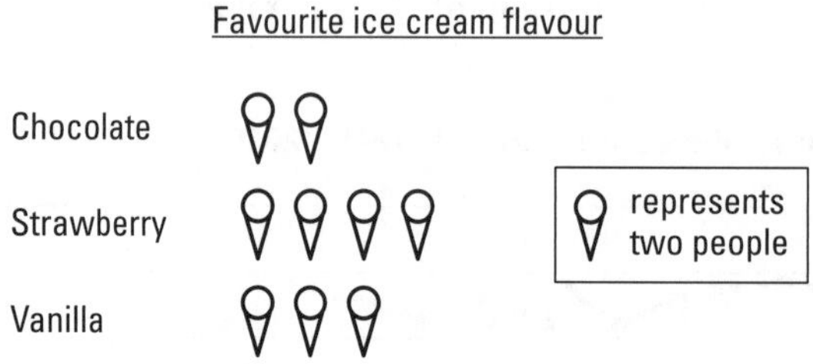

Favourite ice cream flavour

Figure 17-7: A pictogram.

Chocolate   ♀ ♀

Strawberry   ♀ ♀ ♀ ♀

Vanilla   ♀ ♀ ♀

♀ represents two people

Labelling and titling your graphs

Some people are really obsessive about making sure graphs are properly labelled. I'm not one of those people, but sadly the odds are your examiner or your boss is a stickler. For the sake of form, I go through the labelling spiel in this section.

Every graph needs to have a title telling you what the thing shows – for instance, 'CD collection' tells you at a glance what Figure 17-4 is about.

How are height and shoe size related?

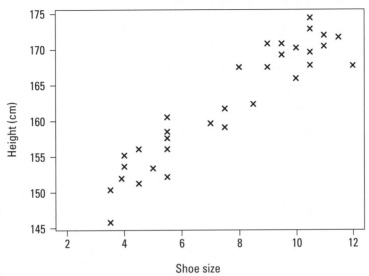

Figure 17-8:
A scatter
plot.

In bar charts and line graphs, you label the *axes* – the horizontal and vertical edges of the graph – so you can see what the graph is measuring. In Figure 17-6, I put 'per cent turnout' on the side – the *y-axis* – and 'year' along the bottom – the *x-axis*. In Figure 17-4, I label each bar to show what it represents; you can write on the bars themselves, or along the horizontal axis.

In a pie chart, try to label the slices clearly, either by writing on them, labelling them with arrows, or using a colour- or shading-coded key so you can look up what each slice represents. You can see a labelled pie chart in Figure 17-9 later on.

Ordering at the bar chart

You normally use a bar chart to show the relative sizes of the categories you care about. A bar chart's one of the easiest graphs to draw – all you do is work out how tall each bar needs to be and draw the bars, and then do a bit of labelling.

In real life, most people draw bar charts and most other types of graph using a computer program. The only time you really need to draw graphs by hand is in maths classes and exams. Jump to the section 'Drawing Graphs on the Computer' for the low-down on computerising your graphs.

Probably the trickiest thing for a bar chart is figuring out how to set the scale so you know how tall to make the boxes. Here's my best strategy:

1. **Find the biggest number you need to plot.**

2. **Repeatedly divide that number by ten until you get a number of centimetres that fits in the vertical space you have available.**

   Make a note of how many divide sums you did. Write down '1' followed by the same number of zeros as you did divide sums. This is your vertical scale, a very important number. Underline it or put a box around the number – you'll need this number when you come to draw the graph.

3. **Draw the axes.**

   Draw a sideways line at the bottom and a vertical line on the left. On the vertical axis, put a little tick-mark every centimetre.

4. **Label the vertical axis.**

   Next to the first tick-mark above the horizontal line, write the number from Step 2. Above that, write 2 followed by the same number of zeros, and so on.

5. **Take each value you want to plot and divide it by the scale number you worked out in Step 2.**

   Your answers will be the heights of your bars – write them down.

6. **Work out how wide to make the bars.**

   Work out how much horizontal space you have available. Divide that number by the number of bars, and then halve your answer. Any width smaller than that will be fine.

7. **Draw the boxes.**

   For each box, draw two vertical straight lines starting at the horizontal line. They should be the length you worked out in Step 4 and the width you worked out in Step 5. Join the two lines at the top. Leave a gap of the same width before doing the same thing for the next bar.

After you finish, you should have your very own miniature Manhattan skyline!

Cooking up a pie chart

To draw a pie chart, you figure out the angles you need to use for your various slices. You can do this using the Table of Joy, which takes an awful lot of the headache out of the whole affair. You check your angles are correct – if the angles don't add up to 360, you have a problem. And then you draw and label the slices of pie.

A multiple-choice exam question can't ask you to draw a pie chart. But the question may ask you which angle is correct. Unless you desperately care about drawing pie charts, you probably only need to read the section 'Finding your angle' for your numeracy exam.

Finding your angle

Working out the angles for each sector is less frightening than you might think, as you can use the Table of Joy (which I introduce in Chapter 8). All you have to remember is that the total of the angles in your pie chart must be 360 degrees. To find out the size of the angle for a slice, here's what you do:

1. **Draw out a Table of Joy noughts-and-crosses grid.**

   Leave plenty of room for labels.

2. **Label the middle column 'number' and the right-hand column 'degrees'.**

3. **Label the middle row 'slice' and the bottom row 'total'.**

4. **Write the value of the category you're interested in in the 'number/ slice' square, the total of the values in the 'number/total' square and 360 in the 'total/degrees' square.**

   Put a question mark in the 'slice/degrees' square.

5. **Write down the Table-of-Joy sum.**

   Times together the two squares touching the question mark (360 × the slice number) divided by the total number.

6. **Work out the sum.**

After you have an angle for each slice, check your angles add up to $360°$ – if not, at least one of the numbers is wrong. Which number is out of step isn't always obvious, and sometimes you have to do the sums all over again. Sorry.

Drawing the chart

To draw a pie chart, you need a compass (the two-pronged thing with a pointy end and a pencil – not the navigational kind), a protractor (the half-circle thing with angles marked on it), and a ruler. Here's what to do:

1. **Use a compass to draw a circle.**

   Mark the centre with a pencil. Give yourself plenty of space – a tiny pie chart is worse than no pie chart.

2. **Draw a line from the centre vertically up.**

   The line doesn't have to be perfect.

3. **Use the protractor to measure the first angle in your list.**

   Check out Chapter 15 if you're not sure how to use your protractor. You can go either way around the circle – for some reason I prefer clockwise.

4. **Mark the angle and place the ruler touching the centre and the mark.**

   Draw a line along the ruler from the centre to the edge of the circle.

5. **Twist the protractor around and measure the next angle, starting from the line you just drew.**

6. **Keep doing this all the way round.**

7. **The last slice should bring you neatly back to where you started.**

   You're nearly done. All you need to do now is label your slices.

Labelling the chart

Here are the most common ways of labelling a pie chart:

- ✔ **Use a separate key:** Number or colour each of the pie sections and draw a box near the graph. In the box, list the colours or the numbers next to the labels they represent.

- ✔ **Label the sectors:** If your slices are large enough, you can simply write the names of the categories in the slices.

- ✔ **Place labels near or in the chart:** Label each sector, as in Figure 17-9.

**Customer satisfaction
survey results**

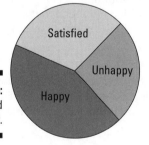

Figure 17-9:
A completed
pie chart.

Forming an orderly line graph

You use a line graph to show how a quantity varies over time. If you look in the business section of a newspaper, you'll probably see line graphs showing how share and currency prices have changed recently. If you visit a busy train station, you may find a line graph showing how a train company has performed over the past few months. Anything that changes over time can reasonably be shown on a line graph.

The main idea of a line graph is that the height of the graph at any given time shows the value of whatever you're measuring at that time.

To get started on drawing a line graph, you need the following bits of equipment:

- ✔ A ruler
- ✔ Graph or squared paper
- ✔ A sharp pencil and a pen

Setting up your axes

The axes show your readers what your graph means. You need to decide which axis is which, work out the scales, draw the axes and then label them.

To work out which axis is which, you need to decide which thing *depends* on what. The thing that is dependent on the other thing goes on the vertical axis. For example, if you want to plot mobile-phone sales over time, the sales don't cause the time to change – instead, the time causes the sales to change. We say that the sales *depend* on the time. Therefore time goes along the bottom, and the sales go on the vertical axis.

Vertical means up and down. Horizontal means 'like the horizon' – going from side to side. Time almost always goes on the horizontal axis.

To work out the vertical scale, you figure out how much space you have available and the largest value you need to plot. To come up with a simple scale that works, divide (very roughly) the largest value by the space you have available. Round this number up to the next round number that's easy to use. For instance, if you have a value around 80, you may decide 100 is easier to work with; if you have a value around 18, you may choose to work with 20. Write this number down. Maybe put a 'V' beside it so you know it's the vertical scale.

Here are the steps for working out the horizontal scale:

1. **Find the difference between the first data point and the last.**

2. **Decide how much space you have horizontally, in centimetres.**

3. **Divide the difference (from Step 1) by the space (from Step 2).**

   Round up to an easy-to-use number, and write this number next to an H, for horizontal.

After you sort the scales, you draw the axes. This is the easy bit: draw a vertical line near the left of the space you have available, and a horizontal line near the bottom. The lines should cross, but only a little bit, like on the left of Figure 17-10.

'Axes' is the plural form of 'axis'. You say 'the horizontal axis' but 'the axes' when you're talking about both of them.

Now you need to label the axes with numbers and a title:

1. **Reading along the horizontal axis, move one centimetre to the right of the other axis and make a small mark.**

2. **Do this again every centimetre until you run out of room.**

3. **Under the first mark, write the number you worked out for H.**

4. **Add the H number to this and write it under the next mark.**

   Using a round number for H makes it easy to add on! Keep adding H and moving to the right until you run out of little marks.

5. **Repeat Steps 1–4 for the vertical axis . . .**

   . . . but work upwards instead of to the right, and with the number you wrote down as V.

The middle of Figure 17-10 shows what your graph should look like. In Figure 17-10 I picked a vertical range up to 80 and a horizontal range up to 40.

Plotting your points

Plotting the points on your graph is a very dull process that requires you to concentrate quite hard. This is one of the reasons computers are brilliant for drawing graphs: they do all of the drudge work for you. Here's how to plot your graph points by hand:

1. **Look at the first data point you want to plot.**

   Find which number corresponds to the horizontal axis.

2. **Find that number on the horizontal axis and make a little mark.**

3. **Find the other number on the vertical axis and make a little mark.**

4. **Very faintly, using a ruler, draw up from the mark on the horizontal axis and then across from the mark on the vertical axis.**

5. **Boldly put a cross where the two lines meet.**

6. **Repeat Steps 1 to 5 for all of the points you want to plot.**

7. **Join the up points with straight lines.**

The right-hand part of Figure 17-10 shows how to plot your points.

Figure 17-10:
A line graph
in various
stages of
completion.

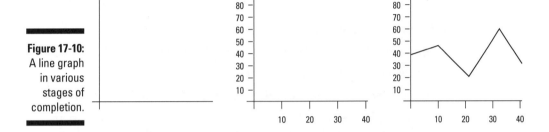

Drawing Graphs on the Computer

In real life, I can think of exactly zero situations in which you need to sit down and draw an accurate graph on paper. As a skill, graph-creating is up there with copperplate handwriting: we have machines that can do that kind of thing far more quickly and accurately than even the most painstaking artist.

These machines are called *computers* – you may have heard of them.

The easiest way to draw a graph using a computer is to use a *spreadsheet* program such as Microsoft Excel or OpenOffice Calc, which is free.

This book isn't a computer instruction manual, and you can find more detailed explanations on how to draw graphs on a computer in *Excel For Dummies*. In this section I simply give you a taste of how to use a spreadsheet to draw the graphs we look at elsewhere in this chapter.

Building virtual bar charts

To draw a bar chart – in fact, any kind of chart – in a spreadsheet, you need to do two things: enter your data into the program, and tell the program to draw a graph.

Computers are both phenomenally smart and phenomenally stupid: they can do amazing feats of calculation in fractions of seconds that would take you or me days of work; but put a bracket in the wrong place and they completely freak out. Entering your data is really important so the computer has every chance of understanding the information.

I assume you have a little bit of computer knowledge in this section: I assume you know how to start up a computer and a program, how to select things, and how to use menus.

Entering your bar chart data

To enter the data for a bar chart, follow these steps in tandem with Figure 17-11:

1. **Open the spreadsheet program.**

2. **In the top-left cell, write the name of your categories.**

   For instance, 'phone company' or 'year'.

3. **In the next cell to the right, write what the bar chart is measuring.**

   For example, 'sales' or 'house prices'. For a multiple-bar chart, write each thing you measure one further cell to the right each time. I then like to make this line bold (try selecting the text and pressing control-B).

4. **Below the top-left cell, write down the actual categories.**

   For the phone companies example, you may have 'Phones'R'Us', 'DummyTel' and so on. For the years example, you may write '2003', '2004', . . .

5. **Below each of the value labels, write the appropriate value for each category.**

   You now have a gorgeous table that your computer understands.

Figure 17-11: A spreadsheet table for a bar chart.

Phone company	Sales
Phones R Us	1243
Dummy Tel	853
UKT	943
BellPhone	732
C2S04	1017

Drawing your bar chart

Here's how to use the table from the previous section to make a beautiful bar chart that you can print out and hang on the wall (if you're so inclined). Here's what you do:

1. **From the 'Insert' menu, select 'Chart'.**

2. **Choose the option 'Bar Chart'.**

3. **Click 'OK' or 'Finish'.**

   Ta-da! You should have a lovely bar chart.

Computers vary slightly. If you don't see precisely the same buttons or menus I talk about, see if your computer offers different but similar options.

Creating virtual pie charts

Drawing a pie chart on the computer is a fantastic experience, at least compared with doing it by hand. It takes seconds, and you don't have to remember how to use a protractor or worry about getting your angles wrong. The main thing is to enter your data properly.

Entering your pie-chart data

To enter the data for a pie chart into your favourite spreadsheet program, follow these steps and look at Figure 17-12 as you go along:

1. **Open the spreadsheet program.**

2. **In the top-left cell, write the name of the categories you have.**

   For example, 'hair colour' or 'client'. I like to select this row and make it bold by pressing control-B.

3. **In the next cell to the right, write what the bar chart is measuring.**

   For example, 'people' or 'income'.

4. **Below the top-left cell, type in the categories.**

   For example, for the hair colour categories you may have 'blonde', 'brown', etc. For the people categories, you may have 'Mr Smith', 'Mr Jones' and so on.

5. **Below each of the value labels, write the appropriate value for each category.**

   Your spreadsheet now has a table ripe for turning into a pie chart.

Figure 17-12: A spreadsheet table for a pie chart.

Hair colour	People
Blonde	8
Brown	12
Red	3
Grey	2
Black	6

Drawing your pie chart

To draw your pie chart, you simply click the right buttons in the right menus. If your computer is a bit different from mine, play about with options and menus that sound similar and see what happens – you won't break the computer.

To draw a pie chart:

1. **Select the data, including the labels.**

2. **From the 'Insert' menu, pick 'Chart . . .'.**

3. **In the dialogue box that appears, click 'Pie Chart'.**

4. **Click 'OK' or 'Finish'.**

   You should see a beautiful pie chart.

Drawing virtual line graphs

In this section I help you draw a line graph using a spreadsheet program. If you've followed the examples in the sections above on bar charts and pie charts on the computer, you'll probably find no surprises in this section. In fact, I bet if you put your mind to it, you could probably figure out how to do a line graph on your own.

Entering your line graph data

To enter the data for a line graph into your favourite spreadsheet program, follow these steps and look at Figure 17-13:

1. **Open the spreadsheet program.**

2. **In the top-left cell, write the name of your independent variable.**

   This is the variable you want on the bottom axis – often 'time'.

3. **In the next cell to the right, write what value the line graph represents – maybe 'sales' or 'temperature'.**

   If you want to draw a multiple-line graph, give all the lines different names.

4. **Below the top-left cell write the values of the independent variable.**

5. **Below each of the value labels, write the appropriate value for each category.**

   This is the value that matches up with the value of the independent variable. Awesome! You now have data the computer will gladly turn into a line graph.

Figure 17-13:
A spreadsheet table
for a line
graph.

Time	Temperature
0	3
3	1
6	0
9	3
12	10
15	11
18	10
21	5

Figure 17-13: A spreadsheet table for a line graph.

Drawing your line graph

Here's how to draw a line graph:

1. **Select the data, including the labels.**

2. **From the 'Insert' menu, pick 'Chart . . .'.**

3. **In the dialogue box that appears, click on 'Line Graph'.**

4. **Click 'OK' or 'Finish'.**

   You should now have a lovely line graph.

Messing around with computer graphs

Right-clicking in various places on a spreadsheet-generated chart or graph gives you a list of options to change. Random right-clicking in a graph can be fun, but finding precisely the right thing to click to make the graph jump through a specific hoop is hard sometimes, even for me.

In my opinion, you have three main approaches to figuring things out on a computer:

- ✔ **Click things at random and see what they do.** This is by far the most effective and fun way to work on a computer. Just remember that Control-Z normally undoes whatever you just did.

- ✔ **Look at the help files.** Pressing F1 for help may tell you something useful, but I've been pressing F1 since I got my first computer in the 1990s and still haven't seen anything useful pop up.

- ✔ **Read a book or ask someone.** I've found a really good series of books to help you with just about anything. The series is called *For Dummies* . . .

Just like with maths, the main thing with computers is not to stress about making mistakes. If something goes horribly wrong, you can easily start again. If you save your work regularly, you can even go back to the last working version. And remember, the world doesn't end if your graph turns pink instead of blue.

Chapter 18

Average Joe

. . . .

In This Chapter
▶ Averaging things out
▶ Making mincemeat of mean, median and mode
▶ Arranging range

. . . .

*A*verage is one of those words with a slightly different meaning in maths than in normal English. When you talk casually, you may make a reference to the 'average man in the street' or say you're 'smarter than the average bear' – average in this context means 'normal' or 'typical'.

In maths, the definition of average is a little more precise, but with the same general idea: you use an average of a set of data (for example, bear intelligence test scores) to determine a typical, normal value.

In this chapter I look at three types of average – the mode, median and mean.

Another statistic (a number that tells you something about some data) you need to know about is the range, or how much difference there is between the top and bottom values. The range tells you how spread out the data are; for example, the mean daily temperature in Canada probably isn't much different from that in the UK, but the range of temperatures is far greater. Also in this chapter I talk about interpreting statistics and using them to make informed choices.

Starting Out with Statistics

A *data set* is a list of observations. A data set can be as simple as information about whether a particular bus is on time or not over the course of a month, or as complicated as a national census, which asks every household in the country a swathe of questions about who lives there. Either way, you have a list of observations – often numbers – that you can do all kinds of interesting maths with, such as finding concise ways to describe the data, and using the descriptions to predict what might happen elsewhere.

A *statistic* is a number that describes a set of data in some way. If your local bus company says '95% of our buses ran on time this week', that's a statistic. If the company says '4,000 journeys made this route', that's also a statistic. Any description of a set of data is a statistic. You can choose from a virtually infinite number of statistics, ranging from the obvious ('How many data points are in the set?' or 'What's the biggest or smallest number?') to the bizarre (the scary-sounding T-tests and Z-scores are only the start of it – but you don't need to know about them unless you study maths further).

Meeting the Three Types of Average

We use three different types of average in maths: the mean, the mode and the median, each of which describes a different 'normal' value. The mean is what you get if you share everything equally, the mode is the most common value, and the median is the value in the middle of a set of data.

Here are some more in-depth definitions:

- ✔ **Median:** In a sense, the median is what you normally mean when you say 'the average man in the street'. The median is the middle-of-the road number – half of the people are above the median and half are below the median. (In America, it's literally the middle of the road: Americans call the central reservation of a highway the 'median'.) If you don't want to think about American highways, try remembering 'medium' clothes are neither large nor small, but somewhere in between. Goldilocks was a median kind of girl.

- ✔ **Mode:** The mode is the most common result. 'Mode' is another word for fashion, so think of it as the most fashionable answer – 'Everyone's learning maths this year!'

- ✔ **Mean:** The mean is what you get by adding up all of the numbers and dividing by how many numbers were in the list. Most people think of the mean when they use the word 'average' in a mathematical sense. In some ways the mean is the fairest average –you get the mean if the numbers are all piled together and then distributed equally. But the mean is also the hardest average to work out.

I know of at least one more average, the mid-range – the mean of the highest and lowest number – but it isn't used very often because it's more volatile than the others. I won't mention it again in this book.

You use the different averages in different situations, depending on what you want to communicate with your sums.

A group of people: Who's the most average?

Imagine five people: Alice, Bob, Charlie, Dave and Eve. Alice and Bob each took an overseas plane trip once last year. Charlie went 16 times, Dave took 8 trips and Eve 4 – as I show in Figure 18-1.

Figure 18-1: Alice, Bob, Charlie, Dave and Eve's travels.

Of these five people, the median number of foreign trips was four – Eve is in the middle of the data because two people took fewer trips and two took more. To work this out, you usually start by arranging the people into order – Alice and Bob with 1 flight each, Eve with 4, Dave with 8 and Charlie with 16.

The mode, though, is one – which in this case also happens to be the smallest value. Two people (Alice and Bob) took one trip each; the other numbers of flights (4, 8 and 16) each had only one person take that many flights – so one flight is the most common number. The five people took a total of 30 trips. If we share out those trips equally between the five people, each would have taken 30 ÷ 5 = 6 trips. Which means the mean is six. Our original set of five data points gave three completely different averages – a median of four, a mode of one and a mean of six.

The man in the middle: The median

The median is the number right in the middle of all the other data – and has the advantage of being relatively easy to work out.

You may read economics reports discussing 'the median income' because the median gives a picture of a much more typical number than the mean does. For the example in the previous section, Charlie's jet-setting ways heavily influenced the mean, but the median would have been the same if Charlie had taken 5 or 160 flights.

It also has the advantage (in my opinion) that it's relatively easy to work out.

Finding the median

To find the median of a set of numbers, you arrange the numbers into order and then find the number exactly in the middle:

1. **If the numbers aren't in order, sort them out.**

   You can arrange them either going up or down.

2. **Circle the number at each end of the list.**

3. **Keep circling numbers two at a time (one from each end) until you have only one or two uncircled numbers.**

4. **If only one number is left, that's the median.**

   You're done!

5. **If two numbers are left, find the mean.**

   Add up the two numbers and divide by two. The answer is the median.

An alternative way of finding the median involves a little more maths but a lot less circling. You find the *median positions*, as in the following recipe:

1. **Put the numbers into order.**

2. **Count how many numbers you have.**

   Then add one.

3. **If this is an even number, divide the number by two.**

   Count this number along from the start: the value in the sorted list you get to is the median.

4. **If your answer to Step 2 is an odd number, divide the number by two.**

   Your number will have .5 at the end. Count this number along from the start, as in Step 3. The .5 means you end up halfway to the next number (for example, the 7.5th value is midway between the seventh and the eighth).

5. **Add the number before where you are to the number after, and halve the result.**

   This is your median.

In a list of five numbers, the median number is the third number. In a list of ten numbers, the median number is midway between the fifth and sixth numbers. The middle number is obvious only if your list has an odd number of items. If your list has an even number of items, you have two points arguing furiously about which of them is in the middle. The only fair way to resolve this dispute is to compromise. The compromise solution is to take the mean of the two values to find the number exactly halfway between them – hence you do the dividing step in each of the methods above when you have an even-length list.

Practising some examples

Say you have the numbers 9, 5, 7, 2, 4, 2, 9, 3, 4. To find the median you put them into order: 2, 2, 3, 4, 4, 5, 7, 9, 9. Notice you repeat the 2s, 4s and 9s because each appears twice in the original list.

To use the first method, repeatedly circle the numbers at each end of the list. That's typographically tricky, so in my example I put brackets around the numbers instead:

(2) 2 3 4 4 5 7 9 (9) – you're down to seven numbers in the middle.

(2) (2) 3 4 4 5 7 (9) (9) – now only five numbers left.

(2) (2) (3) 4 4 5 (7) (9) (9) – almost there – now you have three numbers.

(2) (2) (3) (4) 4 (5) (7) (9) (9) – finally your answer is 4.

To use the second method, you say 'I have nine numbers, so I add one and halve the total (ten) to get five. I want the fifth number in the sorted list, which is 4.'

Here's another example, with a shorter list of numbers: 4, 7, 4, 6, 7, 5. Sorting them into order gives 4, 4, 5, 6, 7, 7. Circling the end numbers gives us the following:

(4) 4 5 6 7 (7)

(4) (4) 5 6 (7) (7)

You have two numbers in the middle, 5 and 6. Add these up and divide by 2 to get 5.5.

Alternatively, you can say 'I have six values, so I add one and halve the total to get 3.5. I want the mean of the third and fourth values, which are 5 and 6 – so the answer is 5.5.'

The more mathematical way is particularly useful when you have a very long list of numbers, as you don't have to write the whole thing out.

Terribly common, darling: The mode

The mode is the most common value in a set of data. I sometimes think the mode is the least useful of the averages, because it's not necessarily in the middle of the data. But then again, sometimes the mode is precisely what you want to describe – for example, 'More people fit this description than any other.'

An example of a useful mode is in a survey of the number of arms people have. A large majority of people have two arms, but some have fewer. Nobody (as far as I know) has three or more arms. The mode in this case is two – the same as the biggest number in the data set.

An example of a less-useful mode is in a survey about winning the lottery. Even though some tickets win prizes, the vast majority of them win nothing at all. If you record the prizes won by ten random tickets, I guarantee you that the mode will be zero – the lowest possible prize.

If you have a list of numbers in order, figuring out which number shows up most often is pretty easy. You simply count the numbers – whichever number you have most of is the mode.

If you have the list 1, 1, 3, 5, 5, 5, 6, 6, 7, 8, 9, 9, 10, you count each number in turn and find you have two 1s, one 3, three 5s, two 6s, one 7, one 8, two 9s and a 10. The number 5 comes up more frequently than any of the others, so the mode of these data is 5.

If the data aren't in a list, I suggest you set up a *tally chart* to help you count the numbers. Here's what to do:

1. **Set up a three-column table like the one in Figure 18-2.**

   Make the middle column as wide as you can.

2. **Write down the first number in the list in the left-hand column.**

   Make a mark next to the number, in the middle column.

3. **Look at the next number in the list.**

   If you haven't already written this number in the left-hand column, write it in a new row and put a mark in the middle column, as in Step 2. If you have written the number before, just put another mark in the middle column next to the number.

   Traditionally, when you make a group of five 1s, you put a cross through them to make them easier to count at the end, but that's up to you.

4. **Go back to Step 3 until you run out of numbers.**

5. **Count up the marks for each number.**

   Whichever number has the most marks by it is the mode.

Try creating your tally chart on squared paper, leaving one square for each 1. You then just look for the longest line of 1s – the number to the left of that line is the mode.

The tally chart in Figure 18-2 is for the list 1, 2, 1, 1, 0, 2, 1, 0, 3, 4, 2, 1, 1, 1, 3, 4, 4. The most common number in the list is 1, which occurs seven times. The mode is 1.

Number	Tally	Count
0		
1		
2		
3		
4		

Figure 18-2: A tally chart before and after you fill it in.

Number	Tally	Count
0	||	2
1	⊬⊬ ||	7
2	|||	3
3	||	2
4	|||	3

Finding the mode in a table of numbers is very easy: whoever made the table has already done the tally chart for you and counted up the 1s. All you do is find the biggest number in the 'count' or 'frequency' column. The number labelling that row is the mode. I give you an example in Figure 18-3.

Figure 18-3: Finding the mode from a table.

Glasses of water	Number of people
0	2
1	5
2	7
3	6
4	1

More people drank two glasses of water than any other number.

The mode of these data is 2.

A mean, mean man

I tell my students that the mean is so-called because it's the meanest thing an examiner can ask. The mean is the hardest of the three averages to work out. But unfortunately the mean is also the most commonly used average. When somebody mentions the average of a set of numbers, more often than not they mean the mean.

Working out the mean involves taking all of the numbers and dividing them equally between the spaces in the list. The mean is a bit like Communism for numbers . . . and in an exam the mean is the kind of question where you can pick up a lot of Marx. Sorry.

Working out the mean of a list of numbers

Here's how to work out the mean of a set of numbers:

1. **Write out a list of all the numbers.**

2. **Add up all the numbers.**

   In a moment, I give you a trick for doing this efficiently.

3. **Count how many numbers are in the list.**

4. **Divide the total from Step 2 by the total in Step 3.**

   The answer is the mean.

5. **Check your answer makes sense.**

   The mean should be somewhere between the highest and lowest numbers in your list.

Adding up a long list of numbers is a chore. In real life you may use a calculator or a spreadsheet. But in an exam you may not have access to either of those helpful devices. I add up longs lists of numbers by working through the list, adding a pair of numbers at a time, and writing the result in the next row. I show you what I mean in Figure 18-4. If you have a number left over at the end of the row, just copy that number into the next row and keep going.

To find the mean of the numbers 1, 2, 1, 1, 0, 2, 1, 0, 3, 4, 2, 1, 1, 1, 3, 4, 4, you add up all the numbers, to get 31. You have 17 numbers, so the mean is $31 \div 17$, which is a shade under 2.

1 + 2	1 + 1	0 + 2	1 + 0	3 + 4	2 + 1	1 + 1	3 + 4	4
3 + 2		2 + 1		7 + 3		2 + 7		4
5 + 3				10 + 9				4
8 + 19								4
27 + 4								
31								

Figure 18-4:
Adding up a
list of
numbers.

Finding the difference between two means

I don't mean to insult your intelligence. I'm sure you know that finding the difference between two means is the same as finding the difference between any other two numbers – you just take the smaller number away from the bigger number. But you may want to know that maths exams often ask questions about the difference of means.

Here are a few things to keep in mind:

- ✔ **Be lazy:** Often the question tells you one of the means, so you don't have to bother working that one out.

- ✔ **Keep things tidy:** Don't mix up your data! Losing track of which data set you're working with is easy. I draw a coloured box around the data I'm working with so I don't get confused.

- ✔ **Count everything carefully:** The data sets may be different sizes and dividing by the wrong number at the end is an easy mistake. Always divide by the number of things in the data set you've added up.

Finding the mean of a frequency table

When an examiner is in a particularly foul mood with the world – his team lost 5–0 at home to Wigan, the chip shop has sold out of sausages and it's been raining all weekend – he may vent his rage and frustration on students like you by asking you to find the mean of a frequency table.

Just because this type of question is seven shades of evil doesn't mean you can't do it. You just need to work through the job carefully and logically. The question involves a lot of sums to do, but none of them is very hard.

The examiner may give you a table that looks something like the one in Figure 18-5 and ask you to find, say, the mean number of cars per house.

Cars per house	Number of households
1	10
2	12
3	7
4	1

Figure 18-5: A frequency table.

The trick is to think about what the table tells you: 10 houses have 1 car each, 12 houses have 2 cars each, 7 houses have 3 cars each and 1 house has 5 cars. You need to think about how many cars and how many houses are in the list.

The first group has 10 cars for 10 houses. The next group has 24 cars for 12 houses – 2 apiece. The third group has 21 cars for 7 houses. The last group has 4 cars in 1 house. That makes 10 + 24 + 21 + 4 = 59 cars for 10 + 12 + 7 + 1 = 30 houses – a mean of 2 cars per household. Phew!

Here are the steps you take:

1. **In each house–car group, times the two numbers together and write them down next to the row.**

2. **Add up the numbers at the end of each row.**

3. **Count up the frequency column – in this case, the houses.**

   When the question asks you for the mean number of 'somethings' per 'something else', the frequency is the 'something else' column, the one that comes after 'per'.

4. **Divide your answer from Step 2 by your answer from Step 3.**

   The answer is the mean.

Don't be tempted to divide your total number (your answer from Step 2) by the number of groups. If you do that, the examiner will laugh bitterly and probably say 'Typical!' Don't give him the satisfaction.

Choosing the right formula

A multiple-choice exam question may ask you which formula to use to work out the mean. Like any formula question, this seems a lot trickier than it really is. All you do is think about the sum you need to do and eliminate the formulas you wouldn't use. Here are some tips:

✔ If the question involves a horrible frequency table, you need a sum that has a bunch of timeses on top and a big add sum on the bottom.

✔ If the question is about a normal mean, you want an add sum on top and a number on the bottom.

✔ Remember that the big bar in a fraction just means 'divide by'.

Home on the Range

The *range* of a set of values is a really useful description of a list to show how spread out the values are – or the difference between the biggest and the smallest values.

You often use a range when comparing two data sets – for example, 'Team A had a slightly lower average than Team B, but Team A's range was far smaller. Therefore Team A is much more reliable.'

In a similar way, you can use range to describe your expected journey time. My record time for driving home is an hour and 50 minutes, but in heavy traffic my journey may take two and a half hours. I say 'My journey to Oxford normally takes two hours, with a range of 40 minutes.' This gives my mum a much better idea of when to put the kettle on.

Calculating range

Working out the range of a set of numbers is pretty straightforward. Here's what you do:

1. **Find the biggest number.**

2. **Find the smallest number.**

3. **Take the smallest away from the biggest.**

   The answer is your range.

The only slightly tricky thing is picking out the biggest and smallest numbers from an unordered list, but putting numbers into order isn't much of a challenge. Alternatively, find a relatively big number in the list that jumps out at you and then go through the list and see whether any other numbers are bigger than it; then you do just the same thing for small numbers.

If you're lucky, a question in your exam may simply say 'Ellen's best golf score was 70 and her worst score was 85. What was the range of her scores?' You take the smaller number (70) away from the bigger number (85) and get a range of 15.

Don't overcomplicate things! Just look for the simplest thing that could possibly work.

More frequently, the examiner gives you a list of numbers and asks you to find the range. Let's use the following list of numbers: 1, 2, 1, 1, 0, 2, 1, 0, 3, 4, 2, 1, 1, 1, 3, 4, 4. The biggest number in the list is 4. The smallest number is 0. The range is 4 − 0 = 4.

Occasionally the examiner gives you a big table of numbers and asks you to find the range of the table. This is no different from finding the range of a list of numbers: you find the biggest and smallest numbers in the table and take them away as usual. For example, the biggest number in Figure 18-6 is 684, and the smallest number is 509. When you take them away, you get 175, which is the range of the data.

Figure 18-6:
The kind of table you might need to find a range for.

Loaves of bread sold				
	Week 1	Week 2	Week 3	Week 4
Monday	547	548	659	584
Wednesday	533	684	509	662
Friday	660	591	667	638

Rightly or wrongly, examiners think of range as a fairly easy calculation, and some try to compensate by making the questions harder. If you have a complicated table, make sure you look at the right values and check you have the highest and lowest numbers.

Pulling range out of a graph

You may need to find the range of data given in a graph. As long as you're comfortable reading values from a graph – flick through Chapter 16 if not – you're all set. Here's what you do:

1. **Find the highest value on the graph.**

   Look for the value of the tallest bar or the highest point on a line graph.

2. **Find the lowest value on the graph.**

   Look for the value of the smallest bar or the lowest point.

3. **Take the value from Step 2 away from the value in Step 1.**

   The answer is your range.

The only difference is you pull the relevant numbers from the graph before you do the taking-away sum. In the example in Figure 18-7, the highest temperature recorded in the graph is 18°C and the lowest temperature is 11°C. The range of these two temperatures is simply 18 – 11 = 7 degrees.

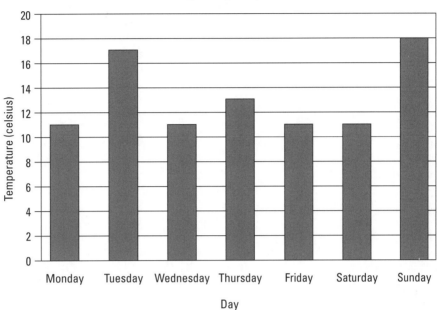

Figure 18-7:
A tempera-
ture graph.

Interpreting your answers

The point of statistics isn't to come up with a number and say 'There – done!' Instead, statistics describe complicated data in a simple but meaningful way, so you can make informed decisions about what to do.

Say you have a train to catch in 10 minutes. You know your drive to the station takes a mean time of 5 minutes, with a range of 15 minutes (some days you just can't park). Meanwhile, walking to the station usually takes 7 minutes, with a range of 2 minutes. On this occasion,

your best option is to walk, even though walking is usually slower. The longest your walk takes is 9 minutes *at most*, but the longest your drive takes is 20 minutes – in which case you miss your train.

This short section may really be beyond the scope of *Basic Maths For Dummies*, but I wanted to show you the kind of thing you can do with statistics. After you figure out what you're doing, statistics can be – whisper it – fun.

Chapter 19

What Are the Chances?

*P*robability is one of those concepts that seem fairly clear-cut on the surface but get murkier and murkier as you get into the details. Schisms exist between different types of probability theorist, and I think they stage regular mock battles in Las Vegas. Probably.

You don't need to pick a side in the dispute at this stage. As far as you care, probability just measures how certain you are that something is going to happen. You're almost certain that the sun is going to rise tomorrow, so you can say that the event – the sun rising tomorrow – has a probability of one (or slightly less if you believe in any of the doomsday stuff).

Likewise, you know Scotland won't win the 2012 football World Cup, because there isn't a world cup that year (and even if there were . . .), so the event – Scotland winning the 2012 football World Cup – has a probability of zero.

Anything that might happen – from *Basic Maths For Dummies* being turned into an international blockbuster movie, to the weather being sunny tomorrow, to the coin I toss coming up heads, to *The Archers* being on Radio 4 – has a probability somewhere between 0 and 1. The higher the number, the more likely the event is.

In this chapter, I take you through the probability you already use in your daily life, how to estimate probabilities by doing experiments, and how to calculate probabilities using the information you already have or can work out.

Understanding basic probability is an important part of everyday maths. Probability isn't currently part of numeracy tests, but understanding there's more to life than 'yes and no' will serve you well.

You probably know this already . . .

You use probabilistic language all the time in your day-to-day life. If someone asks 'Are you coming to the party on Saturday?' you may respond 'Probably' or 'Probably not.' You talk about something being more likely or unlikely. You may call something a million-to-one shot or a 50–50 chance, or say you're 99 per cent sure of something. On the weather forecast, you may hear about a 40 per cent chance of rain.

All of these things are probabilities, although not expressed the way you see probabilities in maths. In maths you give the probability as a number between 0 and 1.

You use probability to describe accurately something in the future that may or may not happen. Using probabilities allows you to say not only 'maybe' or 'this is more likely than that' but also how much more likely one thing is than another.

Probability as a Number

You usually write probability as a number – either a fraction or a decimal – between 0 and 1. The number means the proportion of times you expect the event to happen if you repeat the situation over and over again. Less precisely, the number measures how strongly you believe something will happen.

For example, if you expect something to happen once in every million times you try, the probability is one in a million ($\frac{1}{1,000,000}$). If you expect the event to happen 50 times in every 100 (such as a coin landing heads up), the probability is $\frac{50}{100}$ or $\frac{1}{2}$ or even 0.5. Something you're 99 per cent sure of has a probability of 0.99.

The closer the probability is to 1, the more likely the event is to occur.

Considering certainty and impossibility

Some things are absolutely stone-cold certain. For example, I am *certain* that if I put 2 + 2 into my (working) calculator, the calculator will give me the answer 4. No possible situation exists where I could put 2 + 2 into my calculator and get anything other than 4 as an answer. Mathematically, the probability of this event is one.

Likewise, some things are flat-out *impossible*. I can't attempt the high-jump and inadvertently jump over the moon. I could try this once a minute for the rest of my life and never get near the moon. Mathematically, the probability of me jumping over the moon is zero.

The probability of a certain event is one. The probability of an impossible event is zero. All other probabilities are in between one and zero. If you do a probability sum and get an answer above one or below zero, you've made a mistake somewhere.

Tossing a coin

Some things are equally likely to happen or not happen.

The archetypal 'equally likely' experiment is tossing a fair coin – you expect the coin to land on heads exactly as often as it lands on tails. Other things that are equally likely include throwing an odd or even number on a single die, and removing a black or red card from a full pack of cards. You can probably come up with several other examples.

You may call these things '50–50 shots' – either result comes up 50 per cent of the time, and 50 per cent is half . . . which is the probability of these things.

The probability of an event that either happens or doesn't happen with equal likelihood is 0.5 or ½.

One in . . . whatever

If all you could talk about in probability was coin tosses and even chances, it wouldn't be a very interesting subject. It's easy to extend the idea to any situation where all of the outcomes are equally likely – picking a card out of a pack, drawing a raffle ticket out of a hat, or (more commonly) rolling a die.

Irregular plural alert! I talk about dice quite a bit in this chapter. If you have only one of these objects, the correct word is a *die*. I have absolutely no idea why this is. The English language is a mysterious and awkward thing.

If you roll a die, you don't have a 50–50 chance of throwing a six.

Assuming the die is fair, all six possible outcomes are equally possible, so we split the total probability evenly between the six numbers. You're *certain* the die will land on one of the six numbers (if you throw it properly), so the total probability we have to split between the six numbers is one.

To split something evenly, you need to divide. The sum here is $1 \div 6$, which is $\frac{1}{6}$.

This argument works for anything where the probabilities are equally likely. I'm probably the only geek in the world who doesn't own a 20-sided die, but I know, using the same reasoning, that the probability of rolling the number 20 on a 20-sided die is $\frac{1}{20}$.

Likewise, on a fair four-sided spinner, the probability of a particular segment coming up is $\frac{1}{4}$.

When you have a number of outcomes, each with an equal chance, the probability of each is one divided by the number of outcomes.

Looking at the other side of the coin

If you know the probability of something happening – say, a coin coming up heads, a die landing on six, a roulette ball landing in a specific slot – you can also work out the probability of that thing *not* happening.

You're certain that *something* is going to happen – one of the outcomes has to be the one that comes up – which means that the probabilities of all of the possible outcomes has to add up to one. (This is true in any probability sum; the total probabilities of all the outcomes has to be one). In this case, you've already 'used up' the probability you worked out for the first thing. The probability of your thing *not* happening is what's left over after you take away the original probability from one.

The sum is simply 'one take away the probability'.

Here are some examples:

- The probability of a coin landing on heads is $\frac{1}{2}$. The probability of the coin *not* landing on heads is $(1 - \frac{1}{2}) = \frac{1}{2}$.
- The probability of a die landing on six is $\frac{1}{6}$. The probability of it not landing on six is $(1 - \frac{1}{6}) = \frac{5}{6}$.
- The probability of a roulette ball landing in a specific slot is $\frac{1}{37}$ (on a European wheel, there are slots for the numbers 0 to 36). The probability of you betting on the wrong number is $1 - \frac{1}{37} = \frac{36}{37}$.

Dealing cards

Working out the probability of pulling a face card – jack, queen or king – out of a pack of cards is a bit more complex.

You can work out the probability of pulling each kind of face card individually without too much trouble. A pack of cards has 13 card values (ace to ten, and three face cards), so the probability of pulling any particular card value out is $\frac{1}{13}$. To get the combined probability, you add up the probability for each value, and get $\frac{1}{13}$ (for a jack) + $\frac{1}{13}$ (for a queen) + $\frac{1}{13}$ (for a king), making $\frac{3}{13}$ altogether.

You can only add up the probabilities when events are *mutually exclusive* – no more than one of them can happen. I talk about this in the section 'You Can't Have It Both Ways' later in this chapter.

Practising probability with the number line

Marking where you think a probability lies on a number line (like the one I talk about in Chapter 3) is a more obvious way to deal with probabilities than using numbers. The left end of the line represents 'impossible' and the right end represents 'certain'. The very middle of the number line means a 50–50 chance. The more unlikely an event is, the further to the left of the line you mark the probability. The more you think something will happen, the further to the right the event goes on the number line.

The number line lets you give a gut-feeling indication of how likely something is to happen – you're pretty sure the sun will rise tomorrow, so you can mark the sun rising at the very right end of the number line. You think the supermarket having bananas for sale tomorrow is a bit less likely, so that event goes a little to the left on the number line.

Have a look at Figure 19-1, where I show a number line with some events marked on it.

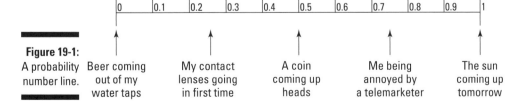

Figure 19-1:
A probability number line.

| 0 | 0.1 | 0.2 | 0.3 | 0.4 | 0.5 | 0.6 | 0.7 | 0.8 | 0.9 | 1 |

Beer coming out of my water taps

My contact lenses going in first time

A coin coming up heads

Me being annoyed by a telemarketer

The sun coming up tomorrow

Converting your gut-feeling estimate on the number line into an actual number is easy if you follow these simple steps:

1. **Measure the number line.**

2. **Measure how far from the left-hand end the mark is.**

3. **Divide your answer from Step 2 by your answer from Step 1.**

Making your probability number line 10 centimetres long will make the divide sum much easier.

Experimenting and Estimating

The most reliable way to estimate the probability of an event happening is to do an experiment. I have a science background, so my answer to everything is 'do an experiment'. I have the scars to prove it. For the purposes of this chapter, though, unless you throw your die really hard and get an unlucky rebound, the experiments are unlikely to injure you.

You need to know about two kinds of probability experiment. In the first type, you repeat things under pretty much the same conditions over and over again. Anything involving cards and dice is this kind of experiment, as are most laboratory experiments.

In the other type of experiment you try to estimate the probability of something even though controlling the conditions is impossible. For example, how likely is My Lovely Horse to win the 3.30 p.m. at Chepstow? What are the chances of it raining tomorrow?

Dice, cards and spinners

Setting up an experiment with dice, cards or spinners is easy but really boring. Here's what you do:

1. **Decide how many *trials* you want to run.**

   How many times will you throw the die or spin the spinner?

2. **Set up a tally chart like the one in Figure 19-2.**

   Head to Chapter 17 if you don't know what a tally chart is.

3. **If you're using cards, shuffle them.**

   This helps you make the outcome as random as possible.

4. **Run a trial!**

   Throw the die or pick a card or spin the spinner, as appropriate.

5. **Record the result on your tally chart.**

6. **Go back to Step 3 over and over until you've completed all of the trials you decided you were going to run in Step 1.**

To work out the probability of your chosen category, you divide the total number of tally marks in your category by the total number of trials. For example, if you want to know the probability of throwing a six on a particular die, you may throw the die 100 times and get 15 sixes. Your probability estimate is $^{15}/_{100}$ or 0.15.

Number	Tally	Count
1	⌧ ⌧ ⌧ |||	17
2	⌧ ⌧ ⌧ ||||	19
3	⌧ ⌧ ⌧ ⌧	20
4	⌧ ⌧ ||	12
5	⌧ ⌧ ⌧ ||	17
6	⌧ ⌧ ⌧	15

Figure 19-2:
A completed tally chart.

How many times . . . ?

You can also work out how many times you expect an event with a given probability to occur if you run the experiment some number of times.

You can do this with the Table of Joy if you want, but I won't show you how – I think the table's slightly overkill for this sum. Check out Chapter 8 if you need the low-down on the Table of Joy.

Instead, you simply times the number of trials by the probability.

If you have the following exam question:

> *A spinner has a probability of 0.2 that it will land on red. If you spun the spinner 200 times, how often would you expect it to land on red?*

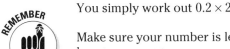

You simply work out $0.2 \times 200 = 40$ times.

Make sure your number is less than the number of trials – or else something has gone wrong.

Best-guess probability

The number that comes out of an experiment like this is really only an *estimate* of the true probability. If you roll six dice at the same time, on average you'll see one 3, but you may also throw none or two or more 3s. Even in a situation where you know the probability accurately – for example for a die you have a ⅙ chance of throwing any given side – the experiment may not give you a precise answer. However, if you repeat your experiment over and over again, your combined results will get closer and closer to the true probability.

You may hear a Met Office official saying something like 'We have a 40 per cent chance of rain in Newcastle tomorrow.' But they can't set up 100 Newcastles with identical weather patterns and see how tomorrow pans out in each of them – and even if they did, they couldn't know whether their number was accurate.

Instead, the Met Office can do two things: first, they set up virtual Newcastles in a sophisticated computer model and run a large number of experiments (very similar to the ones in the previous section) to get their answer. Second, they combine this technique with an equally sophisticated approach called 'educated guesswork' – somebody looks at the weather systems around the UK, the prevailing wind conditions and the weather over the past few weeks and compares the current situation with situations in the past. A team of meteorologists then come together and agree on a sensible forecast.

The moral of this story is some things are too complicated to predict accurately – and if they are, guessing is okay.

Putting Many Things Together

Probability can get very complicated. Much, much more complicated than anything I cover in this chapter. In this section I talk about the probability of *combined events* – two things that happen at once or after each other.

For example, rolling two dice is a combined event – you have two separate events, one for each die. A more complicated combined event involves the probabilities of the weather being sunny tomorrow and me going for a run. Neither of these things is certain, but I'm interested in both of them. (In this case, one affects the other; if the weather's nice, I'm much more likely to take a run. Even though my decision to go for a run doesn't affect the weather, the weather certainly influences my decision! By contrast, the dice don't affect each other at all.)

Most people can't visualise an abstract idea like probability. The number line is very useful for single probabilities, but it breaks down a bit when you have to think about two separate events.

Instead, you normally use a probability tree (the method preferred by most maths teachers, and one of the least intuitive and most confusing methods known to humankind) or a probability table (my preferred method because it's really easy).

For completeness, I also tell you about probability squares in this chapter, which work well for people who like shapes.

The probability table is a little less general than the probability tree, because you can only sensibly deal with two events at a time. As far as you're concerned, though, the probability table covers pretty much every eventuality. Even probability trees get messy if you have more than two events – and sometimes even before that.

Probability trees

Teachers around the world use probability trees. Probability trees are quite versatile if you use them properly, but if I had a pound for every time I heard the question 'Do you add or times as you go across the way?' I'd be a very rich man indeed. (It's times. I show you why in a minute.)

The idea of a probability tree is to start on the left with 'the whole thing' or one. Every time several possible outcomes exist the probability in that branch splits off into a smaller branch for each outcome. I show an example in Figure 19-3 to give you a flavour of it. In this figure, I ask the question: what is the probability of throwing a head and a tail if I toss two coins?

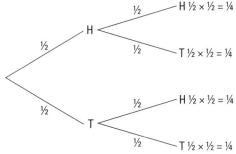

Figure 19-3:
A probability tree about coins.

At the first event – the first coin toss – I have a ½ chance of the coin coming up heads and a ½ chance of the coin coming up tails. On the left of the tree, the line splits into two parts: one with H (for heads) at the end, labelled ½, and another with T (for tails) at the end, also labelled ½.

At the second event – the second coin toss – each of the branches splits again. The outcomes and probabilities are the same as before for each, so we get a branch with H at the end labelled ½ and a branch with T at the end, also labelled ½.

You can use the tree to work out the probability of any combination of events. For example, to find the probability of the first coin being a head and the second a tail, you follow the branch from the first split to H, and then the branch from there to T, and then times the labels (the probabilities) together. You get a half of a half: $½ \times ½ = ¼$.

Timesing by a fraction makes your number smaller. As you go across the tree, you split up the probability at each event. Think of a river splitting up into smaller streams as you head towards the source.

Here are the steps for drawing your own probability tree:

1. **Decide what happens first and how many possible outcomes there are.**

   Draw that many branches coming out of a point on the left of the page. Be sure to give yourself plenty of room for labels.

2. **For each of the outcomes, work out the probability and write it beside the line.**

   Also write something at the end of the line to tell you what the outcome is.

3. **For each branch, repeat the process.**

   Count outcomes, draw lines and label.

4. **Working from the start point to the end of each branch, times together the probabilities of the branches you go along.**

   Write the result at the end of the branch.

5. **Check that the end probabilities add up to one.**

   If not, you've made a mistake somewhere.

You can find the probability of any combination of events you care about by following the branches through the probability tree.

Probability tables

A probability table is a way of representing probabilities. I show an example of a probability table in Figure 19-4, looking at the events 'throwing a head on a fair 50p piece' and 'throwing a head on a fair £1 coin.'

		50p piece		
		Head	Tail	Total
£1 coin	Head			0.5
	Tail			0.5
	Total	0.5	0.5	1

		50p piece		
		Head	Tail	Total
£1 coin	Head	0.25	0.25	0.5
	Tail	0.25	0.25	0.5
	Total	0.5	0.5	1

Figure 19-4: A probability table.

You use the *total* probability of each event to work out the probability of each pair of events. Here's how to draw your probability table:

1. **Count how many possible outcomes the first event has.**

   Your table needs that many columns, plus another two.

2. **Count how many possible outcomes the second event has.**

   Your table needs that many rows, plus another two.

3. **Draw a table with the appropriate number of rows and columns.**

4. **Label the columns.**

   Leave the first column blank and then list the outcomes for the first event, followed by 'total' in the final column.

5. **Label the rows.**

   Leave the first row blank and then list the outcomes for the second event, followed by 'total' in the final row.

6. **The total of all the probabilities has to be one, so put 1 in the 'total/ total' cell in the bottom right.**

7. **Find the probability of one of the events and write it in the 'total' box for that row or column.**

   Do the same for the other event.

8. **Find the probability of a combined event.**

   If you already know this probability, write it in the right box. If not, times the two 'total' numbers you worked out in Step 7 and write the answer in the box in the same row and column as both.

9. **Fill in the remaining boxes.**

   The cells in any row or column have to add up to the 'total' at the end or bottom of it.

Now you can look up the probability of any event by finding the right row and column.

Probability squares

Probability squares are a really good option when you have two *independent* events – two things that don't affect each other (for more on this, see the section 'Independence Day' later in this chapter). You may like probability squares if you're more a 'shapes person' than a 'numbers person'. The probability square extends the probability number line into two dimensions by drawing two separate number lines at right angles – one line for each event. I show what I mean in Figure 19-5.

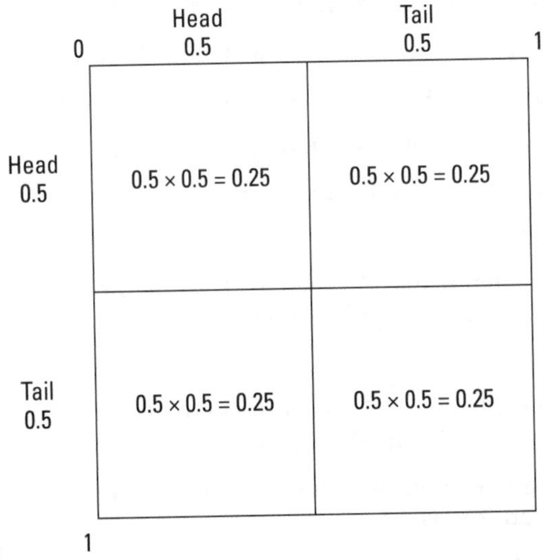

Figure 19-5: A probability square.

For two independent events, here's how to use the probability square:

1. **Draw a rough square.**

   Mark the top-left corner with 0, and the top-right and bottom-left corners both with 1. (Don't bother with the other corner.)

2. **On the horizontal probability line, mark the probability of the first event.**

   You don't have to measure the line. Just put a cross somewhere that looks reasonable.

3. **Draw a line down the page from your cross to the opposite edge of the square.**

4. **Label the horizontal probability line with the probability (to the left of the line) and the 'what's left over' (one take away the probability) on the right.**

5. **On the vertical probability line, mark the probability of the second event.**

   Don't measure – just estimate and draw a small cross.

6. **Draw a line across the square from the small cross.**

   Your diagram should now look like a big cross. The top-left rectangle represents the probability of both events happening; the top-right rectangle shows the probability of the second event happening but not the first; the bottom-left rectangle gives you the probability of the first event happening but not the second; and the bottom-right rectangle gives you the probability of neither event happening.

7. **Label the vertical line with the probability (above the line) and the number left over (below it).**

8. **To find the probability of both things happening – the top-left quarter – you times the number at the top of the column by the number at the left end of the row.**

   Remember, to get the probability of two events occurring, you have to times them together. Visually, this corresponds to the area of the rectangle, so if you were to draw it accurately the biggest areas would represent the largest probabilities.

9. **To find the probability of any combination of events, find the correct area (as described in Step 6) and times the number at the top of the column by the number at the left end.**

As with many of the methods I offer in this book, this one uses a lot of steps to reach a very simple sum at the end. If you prefer to use a different method, I don't mind – as long as you get the right answer!

Independence Day

If you call a politician 'independent', you imply that their actions have absolutely no bearing on the other politicians, and that *their* actions have no bearing on the independent politician.

Exactly the same concept exists in probability: if you throw two dice, the score on one doesn't tell you anything about the score on the other, so throwing a six on one die is *independent* of throwing a six on the other die. In fact, every event only involving the first die is independent of every event only involving the other die.

By contrast, the probability of it raining and the probability of England winning a cricket match are not independent – rainy cricket scores are very different from sunny cricket scores.

To find the probability of two independent events happening, you times their individual probabilities together. This is why the probability square that I describe in the Section 'Putting Many Things Together' works – you find the area of a square, which involves timesing the sides of the square (which represent probabilities).

You can't have it both ways: Either/or events

With some events, if you know that one thing has happened, you know the other hasn't happened. For example, if I know my coin came up tails, I know for certain that it didn't come up heads. If I'm drinking coffee in my local cafe in Poole, I can't simultaneously be running the London Marathon. The technical term *mutually exclusive* describes two things that can't possibly happen together. Mutually exclusive events are important because they make the maths easier. If I want to know the probability of one event or another event happening and I know the two events are mutually exclusive, I simply add their probabilities together.

If we look at the heads/tails example, the probability of me tossing a head is 0.5 and the probability of me tossing a tail is 0.5. The two events can't both happen, so they're mutually exclusive. Therefore, the probability of me tossing either a head or a tail is 0.5 + 0.5 = 1. I'm certain to get one result or the other.

For a more interesting example, we can think about the probability of throwing a five or a six on a die. The probability of throwing a five is ⅙, and the probability of throwing a six is also ⅙. I can't throw both a five and a six on the same die at the same time, so the two events are mutually exclusive. The probability of throwing either a five or a six is ⅙ + ⅙ = ²⁄₆ = ⅓.

If I pick a card (any card), the events 'the card is an ace' and 'the card is a heart' are *not* mutually exclusive, because the card could be the ace of hearts – that is, these two events can both happen at once. The bad news is the sums are more complicated when the events aren't mutually exclusive. The good news is you don't need to know. Unless you're playing poker at an advanced level or go on to do A-level statistics, you'll probably never have to do the sum.

Doing several things at once: Both or all events

If you know two events are independent, you can find the probability of both of them happening by timesing the individual probabilities together. For example, the probability of throwing a double six is ⅟₃₆ – you have a ⅙ chance of throwing a six on the first die, and a ⅙ chance of throwing a six on the second. Timesing them together gives ⅟₃₆.

If you have more than two independent events, you just keep timesing. The probability of starting a game of Yahtzee and throwing five sixes on your first shot is ⅙ × ⅙ × ⅙ × ⅙ × ⅙ or . . . let me get my calculator . . . 1 in 7,776. (That's actually a bit more likely than I guessed.)

Part V
The Part of Tens

'You & your Basic Maths expertise!
A holiday in Moldistan – a marvellous
exchange rate! 2½ million Moldistani
zhugs to the £!!'

In this part . . .

What would a *For Dummies* book be without a Part of Tens? Here are some useful hints, practical tips and sneaky tricks for getting the best out of basic maths. Find out how to keep your head, ways you can remember facts, the traps examiners can set for you, and how to ace an exam. What more could you ask for?

Chapter 20

Ten Ways to Prepare Yourself Before You Start Studying

- -

In This Chapter

▶ Preparing your body

▶ Helping your brain

▶ Developing a study habit

- -

1 f you want to make anything really difficult for yourself, try to do it under stress. Ideally, fill your world with distractions, put yourself under a crazy amount of time pressure, and threaten to punish yourself if you don't do the job perfectly and on schedule.

Deliberately making things hard is no way to learn anything. The calmer you are, the easier maths is to understand and learn. If you can set yourself up in a peaceful, comfortable environment before you even start on your day's work, you can make yourself many times smarter – and enjoy your studies that much more.

That's right: I said *enjoy* your studies. After you get into the maths zone, you start to discover the thrill of finding things out, making your own discoveries, starting to make maths your own. This amazing feeling can be yours. You just have to start with some peace of mind.

In this chapter I give you a set of strategies, tactics and attitudes to adopt to make your studies easier. The easier you make studying for yourself, the better you're going to do, and the more likely you are to tell your friends to go out and buy this book. (Please do that, by the way. It's widely accepted that doing nice things for other people reduces your stress levels dramatically.)

Talking Yourself Up

The way you talk to yourself makes an incredible difference to how you perform. If I tell myself 'I can't dance and it's no fun anyway,' I'm likely to have a miserable and embarrassing time trying to learn to salsa. If I tell myself, 'Well, it's exercise, and nobody's watching me, and making mistakes is perfectly normal,' then suddenly I'm having more fun than I have any right to.

You can do the same with maths. Instead of saying 'Maths is stupid and difficult. I can't do it, I'm not a maths person and my brain doesn't work that way' or anything else that's demoralising or nonsense, try saying, 'I'm smart, and I can figure it out. I have an amazing brain – I just need to train it a bit more.' Best of all, try saying: 'Ah, I got that wrong. That's interesting. I wonder why?'

Sitting Up Straight

I'm writing this slumped backwards on a beanbag, so I'm probably not in the best position to tell you that posture is important for top performance. Like the positive self-talk thing, it sounds a bit like hippy woo-woo claptrap, but sitting up straight can make a big difference to how you fare in your studies.

Your brain speaks fluent body language. If you sit upright, with your shoulders back and head up, your body says 'I'm confident and capable of dealing with anything life throws at me' and your brain says 'Bring it on!' If you crunch up and cower over your desk, your body says 'Don't bother me, I don't want to interact' and your brain says 'Sorry, closed for business.'

If you start to feel a bit overwhelmed with your studies, take a few seconds to notice how you're sitting – try to adjust yourself slightly and adopt a more commanding posture.

Breathing like a Rock Star

In Chapter 2 I give a blow-by-powerful-blow account of how to calm yourself with diaphragmatic breathing.

I can't emphasise enough how helpful this type of breathing is to me when I feel panicky or stressed or even just out of sorts. Taking a few minutes to breathe deeply and get some oxygen into my body has saved more than one day from being a complete write-off for me. Any serious singer does some diaphragmatic breathing before they go on stage, both to overcome stage fright and to expand their lungs for better singing. Coupled with positive self-talk, diaphragmatic breathing is a tremendously powerful weapon against the panic-monsters.

Putting Out the Welcome Mat

A doormat with 'Welcome!' on it is so much of a cliché that it barely even registers these days, but once upon a time the idea was to make visitors feel immediately at home and to stay longer.

If you set up your learning experience to welcome new knowledge, and to make maths feel comfortable, at home and wanted, you'll find that having new knowledge come and stay in your brain becomes a regular experience. And unlike house guests, you never have to change the sheets for knowledge.

As a comparison, if you dread a visit from the in-laws, you won't enjoy the experience – and if you make it unpleasant enough, they probably won't want to come back. Likewise, if you think 'I *have* to learn this' instead of 'I'd really love to know this', your mind tends to push away what you try to learn.

Putting out a doormat for a book would be pretty daft, unless you happen to like the analogy so much that it's helpful. Instead, I recommend just saying to yourself 'I'd like to read this bit' rather than 'I have to wade through this chapter.'

Making Mistakes

When was the last time you got through a day without making a mistake of any description? Where you didn't put your phone down somewhere and miss a call, where you didn't have to say 'sorry' or 'I mean . . .', where you didn't have to press the delete button on your computer?

I'm betting on never.

As with anything else, you will make mistakes when you figure things out in maths. You mess up. I mess up. Stephen Hawking messes up. Making mistakes is nothing to be afraid of or ashamed of. The language you use can really help you out. Try not to mutter 'I got this wrong, so I must be rubbish at maths.' Instead, say: 'I got this wrong – I wonder why? What can I learn from this?'

Working with Your Limits

On occasion – when I'm studying something interesting, or watching a fascinating movie – I can concentrate like a fiend for hours on end. Other times – say, when I'm cleaning – I'm lucky if I can think straight for two minutes.

You may find that you study best in ten-minute bursts scattered throughout the day. Maybe you prefer two 25-minute sessions with a short break between. You may work better in the morning, in the evening or somewhere in between. Whichever suits you is the one to choose.

Try setting a timer and see what happens if you try to concentrate for 10 minutes, 20 minutes and 40 minutes. I do this with writing: I set a timer for 15 minutes and write furiously until it goes ping. A quarter-hour is short enough not to be intimidating, but long enough to get something useful done.

Try several permutations and see what works best for you. Then stick to those limits.

Turning Studying into a Habit

Remembering what's going on is much easier if you do a few minutes of work each day, rather than trying to cram everything into one long session. Even just five minutes of studying a day can make maths significantly easier to understand.

Try to get into a routine and commit to keeping to the schedule you choose. In Chapter 2 I introduce the calendar of crosses, a really effective psychological ploy to convince yourself to keep at your studies, even when you don't feel like working; *especially* when you don't feel like working.

Try to schedule a regular time to hit the books – I understand that fitting studying around work, family and social commitments isn't easy, and it may not be practical to say 'I'm going to do 20 minutes at 7.30 every evening' – but the more regularly you study, the better your brain will get at accepting that this is study time – and it's better to put energy into study than into protesting about it.

Staying Fed and Watered

When I'm hungry or thirsty, I get super-cranky. I'm normally – I think – laid-back, sympathetic and kind, but if my body isn't properly fuelled, I turn into the incredible sulk – and then I'm pretty much useless for anything . . . including maths. If you're not physically comfortable, concentrating on what you're trying to learn is pretty much impossible.

The key is to set yourself up before you start. Grab a snack and a bottle of water or a cuppa. Pop to the loo. Get everything you need in one place. Turn off your phone. Put your music on, if you study with music in the background. Get everything as perfect as you can, as quickly as you can.

The last thing you want is to get halfway through a difficult problem and realise you're ravenous or parched with thirst, so you have to stop and refresh yourself. Getting back into the zone after that is very difficult. The more you set yourself up so that you don't have an excuse to stop studying, the more likely you are to study well.

Getting Your Blood Flowing

Some days, I just don't feel like getting started with whatever tasks I have on my plate for the day. I'll check email, play games, pick up the guitar – do just about anything other than the work I've told myself to do.

When I notice myself getting into this kind of rut, I call myself out of it and stand up. Part of my 'reset' routine is to jump up and down for a few seconds, which has the twin benefits of getting my blood pumping through my veins and annoying my downstairs neighbour. I'm always amazed at how much a little jumping around can wake me up. You can do whatever you like to get your heart-rate up: if you like running on the spot, or dancing, or doing push-ups, or walking up and down the stairs, fine. The important thing is to say 'I don't want to be Mr or Mrs Procrastinator-Pants any more', get the blood flowing, and get back to work.

Don't do serious exercise without a gentle warm-up. You'll hurt yourself. Don't come running to me with a torn hamstring.

Warming Up Gently

Just as you need to warm up before exercising, so the same warning applies to your studies. Trying to dive straight into the hardest thing you have to study is a one-way route to demoralisation, misery and – worst of all – having to study that bit all over again.

Instead, try to start off your study sessions with a few minutes reviewing topics you find a bit easier – preferably ones related to what you want to study. If you can't find a related subject, just check you're okay with your number facts or with fundamental arithmetic to build your confidence.

You're not going to pull a hamstring by studying too hard, too fast, but easing yourself into a session can make the prospect much less daunting.

Chapter 21

Ten Tricks for Remembering Your Number Facts

*P*ersonally, I wish maths tests put much less emphasis on being able to do accurate, rapid-fire mental arithmetic under pressure. I think being able to estimate and feed the right sum into a calculator or computer is far more useful – but at the moment we're stuck with the tests we've got.

I suggest you learn your number facts efficiently and with the minimum of fuss. The quicker you can command total mastery of your number facts, the quicker you can stop having to learn them, and the more time you can spend on the more creative and interesting aspects of maths.

In this chapter I list ten things you can do to make learning your number facts easier.

Playing Games

Many games, online and offline, help you practise and learn your number facts. I include a number of online resources in the nearby sidebar – I'm particularly fond of the BBC Skillswise site and Manga High, because many of the alternatives are a bit patronising.

Playing games online

Dozens of good websites offer games to help you learn maths facts and techniques. Some are fun, others serious, and others somewhere in between. Here are some of my favourites:

✔ www.mangahigh.com – Manga High is a beautifully designed maths games and lessons website. It covers a wide range of topics and is probably my favourite maths site.

✔ www.bbc.co.uk/skillswise/ numbers/ – The BBC Skillswise site isn't quite so polished, but is more focused on adult numeracy. It's organised with factsheets, worksheets and games to keep you on top of the topics you need to learn.

✔ www.mymaths.co.uk – My Maths is a subscription site that concentrates more on GCSE and A-level than adult numeracy. All the same, there are some free sample games that will allow you to practise a few of your skills.

✔ www.mathmotorway.com – Math Motorway is simple but brilliant: answer questions on adding, taking away, dividing and/or multiplying as quickly as you can. Every time you get one right, your car moves forward – can you win the race?

✔ www.counton.org – There are some more nice basic maths games here.

✔ www.dilan4.com/maths/ countdown.htm – The numbers game from Countdown is a great way to sharpen your mental arithmetic skills.

Games are a really useful way of learning anything at all, because they take the chore element out of learning and turn it into something a bit more fun. Playing a number-facts game with cards or on the computer is much less like hard work than writing down endless lists of sums. I strongly recommend games as a learning tool – let me know if you find any good ones.

Flashing Cards

Flash cards aren't very common in the UK, but in America pretty much every successful student spends hours of their revision time writing down the key facts they want to remember on index cards and repeatedly testing themselves until they have all their facts down pat. Using flash cards like this is a bit tedious but *very* effective.

You don't need to go out and buy stacks of index cards – just cut up paper or cardboard into smaller bits and write on those instead.

The traditional way to set up flash cards is to write a question on one side and the answer on the other. Then grab some cards from your 'to learn' stack and go through, reading out the question and answering aloud. Check the back of the card: if you got the question right you put the card to one side; if you're not happy with your answer, put the card to the back of the pile. Keep on answering until you run out of questions.

The more you do this, the quicker and more accurate you get.

Sticking Stickies

At university, many people were frightened to go into my room near exam time, and not just because it smelt of damp laundry. When I was revising, I would cover every available surface with sticky notes covered with details I was meant to know for my exams, and sheets of paper full of colourful equations, diagrams, mnemonics and more.

I was pretty mean when it came to buying adhesive, so anyone who opened my door or moved a muscle a few metres away from my room would cause all of the paper to rustle and fall down, at which point I would yell at them to be more careful.

I didn't have many friends in those days.

 A more moderate version of this technique that doesn't make you the villain of your household is to put five key things you want to remember on sticky notes and put the notes somewhere you're bound to see them – say, on the bottom of your computer screen, the microwave or the bathroom mirror – anything you look at more than once a day and is a suitable surface for sticky notes is a great choice. When you can remember the info on the sticky note without even trying, replace the sticky with a new note covering something else you need to learn.

Counting on Your Fingers

Fingers are the reason we count things in tens, so using your digits to figure out questions is perfectly natural. The disadvantage is that counting on your fingers can be a lot slower than just remembering the facts – and especially in an exam, time isn't something you have a lot of.

I'm a big advocate of saying the sum and the answer aloud after you've worked it out on your fingers – for instance, after you add 6 to 9 to get 15, say 'Nine add 6 is 15.' Repeating the sum aloud convinces your brain that the sum is an *important thing to remember*, just like when you say to yourself 'Don't forget the bread' over and over again when you forget your shopping list. Or is that just me?

You can also use your fingers to keep track of your times tables. I suggest you say out loud 'One six is six, two sixes are twelve' and so on, even though this is slower than counting briskly through '6, 12, 18, 24 . . .' If you're looking for the answer to 7×6, you can easily get confused if you try to keep track of the six times table – tapping your fingers as you go through saves you wondering how far you've got.

Tricking out the Nines

In Chapter 4, I show you how to figure out your nine times table using your fingers, but that's not the only way to work with nines. Here's another way to do nine times anything (up to ten):

1. **Take one away from the number you're timesing by.**

   Write down the answer.

2. **Take your answer to Step 1 away from nine.**

   Write this new answer to the right of your first answer – and there's your answer. For example, to do 7×9, take 1 away from 7 and write down 6. Now take 6 from 9 to get 3, and write that to the right – you get 63, which is 7×9.

A good way to check your answer to a nine times table (up to ten) question is to notice that the two numbers in the answer always add up to nine. For example, 7×9 is 63, and $6 + 3$ is 9. After ten, things get a bit trickier: for example, $11 \times 9 = 99$, and $9 + 9$ is 18 . . . but if you add $1 + 8$, you then get 9. You just have to keep going until you get to a single digit.

This only works for the nine times table. Numbers in the eight times table don't always add up to eight, sadly.

Tricking Out the Other Big Numbers

In Chapter 4, I show you how to use your hands for the more difficult, bottom-right part of the times tables – 6×6 and beyond. Here are some clever tricks you can use to figure out your five, six, seven and eight times tables.

Tricks of six

Here I give you a way to work out your six times table where all you need to know is how to double and treble things (times them by three).

To times a number by six:

1. Times the number by three.
2. Double the answer.

For example, to work out 9×6, do $9 \times 3 = 27$, and then double your answer to get 54. It doesn't matter which way around you do the doubling and trebling, as long as you do them both.

Straight to eight

In this section I show you how to work out your eight times table where all you need to know is how to double things. To times a number by eight:

1. Double the number.
2. Double the answer (this is now four times the original number).
3. Double the answer again.

For example, to do 7×8, double 7 to get 14. Double 14 to get 28. Then double 28 to get 56.

What about seven?

After you know how the sixes and eights work, you have two quick ways to work with sevens. To times a number by seven using the sixes:

1. Work out six times your number.
2. Add the original number to that.

For example, to work out 7×7, do $7 \times 6 = 42$, and add 7 to get 49.

Alternatively, you can times a number by seven using the eights:

1. Work out eight times your number.
2. Take away the original number.

So, to work out 7×7, do $7 \times 8 = 56$ and take away 7 to get 49.

Five alive!

The five times table also has an easy trick: you can times the number by ten and then divide the answer by two, or halve the number first and then times by ten. So, to work out 8×5, do $8 \times 10 = 80$ and divide by 2 to get 40; or, divide by 2 to get 4 and do 4×10 – again, the answer's 40.

If you don't like messing around with fractions, I suggest you times by ten first rather than halving.

Breaking Down and Building Up

Division tricks are pretty much the opposite of the multiplication tricks I describe in the previous section. Don't be surprised – after all, timesing and dividing are the opposite of each other.

Eight: halving over and over

If you want to divide by eight, but doing 'proper' division bothers you, try the following method:

1. Divide the number by two.

2. Divide by two again (you've now divided by $2 \times 2 = 4$).

3. Now divide by two again (you've now divided by $4 \times 2 = 8$).

So, to do $72 \div 8$, halve 72 to get 36, halve again to get 18, and halve one more time to get 9. And that's right: $72 \div 8 = 9$.

Surprised by six

So, would you care to make a prediction about another way of dividing by six?

1. Divide by three.

2. Then divide your answer by two.

You can do these steps in either order. For example, to do $42 \div 6$, you can halve 42 to get 21 and then divide by 3 to get 7. Or you can do $42 \div 3 = 14$ first and then halve the answer to get 7.

Nailing nine

Dividing by nine is just as simple:

1. Divide by three.
2. Divide by three again.

So, faced with $81 \div 9$, you can work out $81 \div 3 = 27$, and then $27 \div 3 = 9$, which is the right answer.

Finally fives

For the five times table, you can double and then divide by ten, or you can divide by ten and then double. So, to work out $85 \div 5$, you double 85 to get 170 and then divide by 10 to get 17. Or you do $85 \div 10 = 8.5$ and then double it to get 17. Doubling rather than dividing first tends to avoid the need for fractions.

Learning from Your Mistakes

To get good at anything, you have to go through a stage of feeling like you're really bad at it. And the way to get through that stage as fast as possible is to make a point of learning from your mistakes.

Whenever you make a mistake in doing a sum, take a moment to note down what you should've done. Then add this to your list of things to learn.

Go through this list as often as you possibly can; the more you practise the things you want to learn, the more quickly you'll find them sticking in your mind.

Learning from your mistakes might sound boring but it pays off really quickly.

Working from What You Know

As I show you in the section 'Tricking out the Other Big Numbers', you can use the next times table up or down to work out the times table you want.

But wait – there's more!

You can also split up any times sum into smaller times sums. For instance, if you want to work out 12×12, there are several ways to split up the sum:

- ✔ If you know 12 is 2×6, you can do $12 \times 2 = 24$ and times that by 6 to get 144.
- ✔ If you know 12 is $10 + 2$, you can do $10 \times 12 = 120$ and $2 \times 12 = 24$ and add the two answers together to make 144.
- ✔ If you're feeling perversely smart and know that 12 is $20 - 8$, you can do $20 \times 12 = 240$, take away $12 \times 8 = 96$ and get 144. I don't recommend that one – but it does work.

You can also split up many divide sums into smaller, easier ones. You can't use all the same tricks as you use for timesing, but here's a division recipe you can try:

1. **Look at the number you're trying to divide by and try to find two numbers that times together to make that number.**

   For example, if you want to divide by 12, you could pick either 2 and 6, or 3 and 4.

2. **Divide by each of those numbers in turn.**

   It doesn't matter which way round, as long as you use both of them.

   So, to divide 576 by 12, you can divide 576 by 2 to get 288, and divide 288 by 6 to get 48. If you prefer, you can divide 576 by 3 to get 192 and divide 192 by 4 to get 48.

This recipe doesn't work for all numbers, but it's worth trying for any number bigger than ten.

Training Yourself with Treats

Although the human brain is a marvellous and sophisticated piece of wetware, far more powerful than a supercomputer, it's also not all that dissimilar from a dog's brain.

If you want to train a dog to learn a behaviour, every time the dog does the task correctly you give the creature a treat. The dog learns to associate sitting on command with the pleasure it gets from a reward, and the next time you say 'Sit!' it is more likely to do as it's told. Good dog.

Your mind works in much the same way: if you reward yourself with a treat after you learn to perform a task well, you do better at repeating your feat at a later time.

I recommend chocolate for humans as a treat for doing well – every time you beat your flashcards record or get a long-division sum right, give yourself a pat on the back and feel the pleasure of having made progress. You've earned it.

Chapter 22

Ten Pitfalls to Avoid

*W*hen you're under pressure in a maths test, you're bound to make some mistakes. But some mistakes are avoidable when you know how; there are also some mistakes you can learn to check for and put right. In this chapter I list the top ten most common pitfalls in maths tests and explain how to avoid them.

Taking Care with Your Calculator

I had a student once who read a question, merrily tapped the numbers into his calculator, and confidently informed me that Mars was five centimetres away. He refused any argument to the contrary on the grounds that his calculator was always right.

Don't be like him.

A machine is only as good as what you tell it to do. Calculators can be amazingly helpful when you use them effectively, but I recommend you always work out rough answers for each step of your calculation so you can check you're doing the right thing along the way.

You're Out of Line!

When you draw a line graph, you plot a series of points on graph paper and join up the dots. Normally you get a smooth or at least smooth-ish curve.

But sometimes you go to draw the line and think, 'Hang on! That point's miles away from the others.' Check that dodgy point very carefully – it's not necessarily wrong, but be suspicious of it.

Making Sure Your Answer Makes Sense

In almost every sum you do, ask yourself whether your answer looks plausible. Here are a few questions to ask about your answer:

- ✔ **Adding:** Is your answer bigger than what you started with?
- ✔ **Taking away:** Is your answer smaller than what you started with?
- ✔ **Real-life problems:** Roughly what would you expect if you just guessed?
- ✔ **Probability:** Is your answer between 0 and 1?

You can probably come up with dozens of similar checks. Before you even begin a question, try to think of as many criteria as possible that your answer has to satisfy.

Distinguishing 'More Than' and 'At Least'

This is one that everyone trips up on at least once. If I say 'I have to pack at least a dozen pairs of socks', then 12 pairs is a perfectly acceptable number. If I say 'I have to take more than a dozen pairs', then 12 is no longer good. I have to take *more than* 12, so 13 is the smallest number I can take.

(I suppose I could take 12 pairs and an odd sock. But that's beside the point.)

Reading the Question

I'm going to be blunt: if you don't read the question correctly, you get the wrong answer. You may be tempted to rush through and do the first thing that comes into your head – even I do that sometimes. And when I do, I mess up.

Take a breath, write down all the information you have, and ponder what the question is asking. Start calculating only after you're fairly certain what you need to do.

Fathoming the Phantom Forty Minutes

Almost every unit you use in maths and science works on powers of 10: 100 centimetres make a metre, 1,000 grams make a kilogram, and so on. The only real exception is time, where 60 seconds are the order of the day. When you try to do normal maths on time, you can end up 'missing' the 40-minute gap between the 60 minutes in an hour and the 100 whatevers that you use in other sums. Check out Chapter 10 on time for more details on avoiding this gigantic elephant trap.

Getting the Wrong Percentage

A couple of things commonly go wrong when you do percentages: you work out the percentage of the wrong thing, or you add when you're meant to take away.

You can get around the 'wrong thing' problem by filling in the Table of Joy that I show you in Chapter 9. To deal with the 'add or take away' problem, simply read the question carefully to see whether your answer needs to be higher or lower than the original number in the question.

Rounding Too Early

If you want to find an answer correct to, say, two decimal places, you may be tempted to round everything as you go along. This is normally okay in day-to-day life, but in a maths test it costs you some accuracy and leaves you with a slightly wrong answer. Instead, wait until the very end to round off your numbers. If you're only working out an approximation, round early and often. See Chapter 5 to discover more about rounding.

Mixing Up the Mean, Mode and Median

'Mode' is another word for fashion. In maths, think of the mode of a set of numbers as the most fashionable, the most popular – the number that comes up most often.

In the USA, the 'median strip' of a road is the central reservation, in the middle of a highway. In maths, the median is the number in the middle of a set of numbers. If you don't care about the American highway system, in itself a fascinating topic, try remembering that median sounds a bit like 'medium', which means 'middling'.

The 'mean' is the meanest thing an examiner can ask: you have to add up everything in a list and then divide by the number of things in the list.

Forgetting to Convert

In 1999, one of NASA's Mars Orbiters disintegrated as it descended through the Martian atmosphere, at a cost of hundreds of millions of dollars.

The cause was traced back to a software error: one of the control programs used imperial units such as inches, and another program used scientific units such as metres. The two programs didn't understand each other's numbers and the mission went catastrophically wrong.

Oopsie.

We can take two lessons from this story:

✔ Whatever happens in your exam, you won't make a spaceship blow up.

✔ Even rocket science isn't rocket science.

Chapter 23

Ten Ways to Make Any Exam Easier

*E*xams can be stressful and frustrating, and not many people enjoy doing them. But jumping through the hoops of an exam doesn't have to be hard. If you can reach a point where you feel prepared, have a good, solid plan, and know how to relax, exams should hold no fear for you. In this chapter I give you some tips to get past the worst of the stress, prepare effectively, work well in the exam, and then come out of the exam smiling like a Cheshire cat.

Know What You're Up Against

Practising on past papers is one of the most effective ways to prepare for an exam. Past papers give you a great idea of the kinds of questions that come up. Go through some exams from previous years – under your own self-imposed exam conditions if you want – and find out which bits you spend the most time on and where you can improve your understanding. Then focus on those areas for your next few study sessions before you try working on another past paper.

Most exam boards let you download past papers from their websites, either free of charge or for a small fee. Some bookshops also sell books of past papers. Or try asking your college tutor if they have any papers you can practise on.

Practise the Hard Parts

I remember watching *Record Breakers* as a kid and being completely baffled by footage of athlete Kris Akabusi dragging a tyre behind him as he ran. I realise now the idea went like this: if you're used to running with a tyre attached to you, when you enter a race without the tyre you can run much quicker than before.

I don't recommend revising with a tyre tied to your back. Instead, try practising slightly harder questions than those you expect to see in the exam, so when you sit the real thing you think 'Wow, this is easy!' Don't beat yourself up over what you can't do. Just see what you can figure out and applaud yourself for questions you answer correctly.

Remember the Basics

Have you ever watched a football team train? Players spend hours practising short passes, turn-and-sprint, keep-ball and other routines that they've done since they could walk. The players haven't forgotten how to run – they're just practising what they spend most of their time in a match doing. One of my coaches had the mantra 'Good players do the simple things well' – and I apply the same mantra to students.

I'm always astonished when top A-level students say things like 'I never understood long division' or 'I haven't really looked at decimals for years.' These students are telling me they haven't practised the fundamentals of maths for so long that they've forgotten them.

Don't be like the students I describe above. Try warming up for your study sessions with something you find easy and will use over and over – perhaps some times tables or estimating exercises. Keep your hand in with the basics and you'll find the more complicated topics easier.

Use the Final Few Minutes before Your Exam

If you tend to forget simple things in exams, the *crib sheet* is your friend. On a piece of paper, write a few key points you need to remember. Then spend the last few minutes before your exam reading the crib sheet over and over again.

You can't take the crib sheet into the exam room, but you can make notes. So, as soon as the examiner tells you to start writing, write down as much of the crib sheet as you can remember.

Make your crib sheet colourful and full of pictures. Most people's brains are better at remembering pictures than lists of information.

Don't Exhaust Yourself

In every exam hall, you'll always see at least one all-nighter zombie – someone who looks like death warmed up and then put back in the fridge, with bags under their eyes and hair like something out of a Tim Burton movie. You can tell at a glance two things about this person: one, they've been up all night studying; and two, they won't do well in the exam, even with their eyes propped open.

Your brain needs sleep to function properly. There aren't many things I think are plain stupid, but pulling an all-nighter before an exam is up there with driving while drunk and eating peanut butter. You can't expect to do high-level intelligent tasks if your brain says 'I must sleep!'

Before your exam, have a good breakfast and drink enough fluids. Your brain runs on this stuff. Trying to think when you're hungry, thirsty or tired is like driving on fumes with no oil in the engine and no air in the tyres. No good can come of it.

Think Positive, or 'I'll Show Me!'

One of my biggest tricks for overcoming anxiety attacks, when my brain tells me 'You're a loser, you can't do that, you'll mess up and everyone will laugh at you,' is to stand up straight, take a deep breath and say 'Right! I'll show me!'

Your brain can be a bully – but it's a bully with no substance behind it. If you assert yourself and say 'Oh yes I can!' or 'Get out of my way, brain!', you can overcome the self-doubt and low confidence that plague most people at some point.

Tell yourself 'I'm smart, I'm capable, and I'm going to show me what I can do.' You may be surprised how much better you do than when you listen to your inner bully.

I've 'shown me' in all kinds of situations – from playing guitar on stage, to making an awkward phone call, to putting bad jokes in my thesis – and the technique's served me very well. The more you train yourself to ignore your inner bully, the less negative influence it has on your life.

Have a Ritual

You may have seen rugby player Jonny Wilkinson take a penalty kick. He walks up to the ball, counts his steps back and to the side, and then does a funny thing with his hands while he focuses. Then he steps up and whacks the ball through the posts like a machine.

I'm not prattling on about rugby here so much as describing the value of a ritual. When Wilkinson takes a kick, he follows a set routine that he's practised over and over again. The routine prepares him mentally for the task he has to do.

I always do the same thing in exams: I open the paper, take a deep breath, tell myself to do well, and quickly read through the paper before starting the questions. This way, I know how every exam starts. I know what happens in the first two minutes and that nothing bad happens in those two minutes. Having a starting ritual removes much of the worry and stress. You can create your own ritual by thinking about *exactly* what you'll do when you sit down for the exam, and in what order. It doesn't have to be complicated – something as simple as 'Take a deep breath, imagine how great I'll feel when I get my results, and then start reading the first question' would work well.

Manage Your Time

In this chapter, the phrase 'manage your time' may be the tip that gains you the most marks. The idea is not to spend too long on one question – if the answer doesn't come out quickly, mark the question with a star and come back to it later.

You have limited time in the exam, so I suggest you spend that time picking up marks you can definitely get before you spend time on marks you may get eventually. Getting to the end of the exam and finding you've missed three easy questions because you were looking at one hard one is a calamity.

You probably know the feeling of thinking really hard about something for ages, and then giving up, only for the answer to hit you halfway through your walk round the park later on. Leaving a question and coming back to it later can be a really efficient exam technique: as you work on the easier questions, your brain can still work on the harder questions in the background.

Guess If You Need To

If time's running out and you've got a minute to answer the last five multiple-choice questions, you don't really have time to read the questions, let alone work out the answers. In this situation you have two possible approaches:

- ✔ Miss out the questions and get a guaranteed zero for those questions.
- ✔ Guess the answers and maybe pick up a few points.

Guess which of these two approaches I recommend? The clue's in the heading.

Numeracy tests usually aren't negatively marked, so you don't lose points for giving a wrong answer. If you guess when you don't have time or are genuinely stuck, the worst that can happen is that you score no marks for that question.

I have a friend who once aced an economics exam by answering the first question with answer 'a', the second 'b', the third 'c' and so on. Another friend swears by picking answer 'b' every time he doesn't know the answer to a multiple-choice question, as he thinks 'b' answers come up more often than the other letters. (I don't think it does . . . but his is no worse a strategy than any other.)

Try to make guessing a last resort – and always check first to see whether you can eliminate any of the answers.

Index

• S •

Notes

Notes

FOR DUMMIES®

Making Everything Easier! ™

UK editions

BUSINESS

Bookkeeping For Dummies
978-0-470-97626-5

Leadership For Dummies
978-0-470-97211-3

Project Management For Dummies
978-0-470-71119-4

REFERENCE

British Politics For Dummies
978-0-470-68637-9

DIY For Dummies
978-0-470-97450-6

Researching Your Family History Online For Dummies
978-0-470-74535-9

HOBBIES

Growing Your Own Fruit & Veg For Dummies
978-0-470-69960-7

Allotment Gardening For Dummies
978-0-470-68641-6

Electronics For Dummies
978-0-470-68178-7

Asperger's Syndrome For Dummies
978-0-470-66087-4

Boosting Self-Esteem For Dummies
978-0-470-74193-1

British Sign Language
For Dummies
978-0-470-69477-0

Coaching with NLP For Dummies
978-0-470-97226-7

Cricket For Dummies
978-0-470-03454-5

Diabetes For Dummies, 3rd Edition
978-0-470-97711-8

English Grammar For Dummies
978-0-470-05752-0

Flirting For Dummies
978-0-470-74259-4

Football For Dummies
978-0-470-68837-3

IBS For Dummies
978-0-470-51737-6

Improving Your Relationship
For Dummies
978-0-470-68472-6

Lean Six Sigma For Dummies
978-0-470-75626-3

Life Coaching For Dummies,
2nd Edition
978-0-470-66554-1

Management For Dummies,
2nd Edition
978-0-470-97769-9

Nutrition For Dummies, 2nd Edition
978-0-470-97276-2

Available wherever books are sold. For more information or to order direct go to www.wiley.com or call +44 (0) 1243 843291

30093 (p1)

FOR DUMMIES®

A world of resources to help you grow

UK editions

SELF–HELP

978-0-470-66541-1

978-0-470-66543-5

978-0-470-66086-7

STUDENTS

978-0-470-68820-5

978-0-470-74711-7

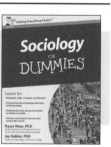

978-1-119-99134-2

HISTORY

978-0-470-68792-5

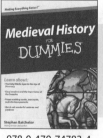

978-0-470-74783-4

978-0-470-97819-1

Origami Kit For Dummies
978-0-470-75857-1

Overcoming Depression For Dummies
978-0-470-69430-5

Positive Psychology For Dummies
978-0-470-72136-0

PRINCE2 For Dummies, 2009 Edition
978-0-470-71025-8

Psychometric Tests For Dummies
978-0-470-75366-8

Reading the Financial Pages
For Dummies
978-0-470-71432-4

Rugby Union For Dummies, 3rd Edition
978-1-119-99092-5

Sage 50 Accounts For Dummies
978-0-470-71558-1

Self-Hypnosis For Dummies
978-0-470-66073-7

Starting a Business For Dummies,
2nd Edition
978-0-470-51806-9

Study Skills For Dummies
978-0-470-74047-7

Teaching English as a Foreign Language
For Dummies
978-0-470-74576-2

Time Management For Dummies
978-0-470-77765-7

Training Your Brain For Dummies
978-0-470-97449-0

Work-Life Balance For Dummies
978-0-470-71380-8

Writing a Dissertation For Dummies
978-0-470-74270-9

30093 (p2)